Aviation in Peace and War

by Sir Frederick Hugh Sykes

CONTENTS

CONCLUSION

INTRODUCTION

Since the earliest communities of human beings first struggled for supremacy and protection, the principles of warfare have remained unchanged. New methods have been evolved and adopted with the progress of science, but no discovery, save perhaps that of gunpowder, has done so much in so short a time to revolutionize the conduct of war as aviation, the youngest, yet destined perhaps to be the most effective fighting-arm. Yet to-day we are only on the threshold of our knowledge, and, striking as was the impetus given to every branch of aeronautics during the four years of war, its future power can only dimly be seen.

We may indeed feel anxious about this great addition of aviation to the destructive power of modern scientific warfare. Bearing its terrors in mind, we may even impotently seek to check its advance, but the appeal of flying is too deep, its elimination is now impossible, and granted that war is inevitable, it must be accepted for good or ill. Fortunately, although with the other great scientific additions, chemical warfare and the submarine, its potentialities for destruction are very great, yet aircraft, unlike the submarine, can be utilized not only in the conduct of war but in the interests of peace, and it is here that we can guide and strengthen it for good. Just as the naval supremacy of Britain was won because commercially we were the greatest seafaring people in the world, so will air supremacy be achieved by that country which, making aviation a part of its everyday life, becomes an airfaring community.

Our nation as a whole has been educated, owing to its geographical situation and by tradition, to interest itself in the broader aspects of marine policy and development. It requires to take the same interest in aviation, a comparatively new subject, unhampered to a great extent by preconceived notions and therefore offering greater scope for individual thought.

The following sketch[1] has been written in the hope that some of those who read it may be inspired to study aviation in one or other of its branches, whether from the historical, technical, strategical, or commercial point of view. Any opinions expressed are, of course, my own and not official.

[1] First written and delivered as the Lees-Knowles Lectures at Cambridge University in February and March, 1921.

I propose first briefly to trace the history of aviation from its beginnings to the outbreak of war; next to describe the evolution of aircraft and of air strategy during the war; and last to estimate the present position and to look into the future.

CHAPTER I

PRE-WAR

EARLY THOUGHTS ON FLIGHT.

The story of the growth of aviation may be likened to that of the discovery and opening up of a new continent. A myth arises, whence no one can tell, of the existence of a new land across the seas. Eventually this land is found without any realization of the importance of the discovery. Then comes the period of colonization and increasing knowledge. But the interior remains unexplored. So, in the case of aviation, man was long convinced, for no scientific reason, that flight was possible. With the first ascent by balloon came the imagined mastery of the air; later, the invention of flight that can be controlled at will. To-day we are still in the stage of colonization. The future resources of the air remain hidden from our view.

The Daedalean myth and the ancient conception of the winged angelic host show how the human mind has long been fascinated by the idea of flight, but the first design of an apparatus to lift man into the air, a parachute-like contrivance, was only reached at the end of the fifteenth century in one of

Leonardo da Vinci's manuscripts. About the same time lived the first of the long line of daring practical aviators, without whom success would never have been achieved, one John Damian, a physician of the Court of James IV of Scotland, who "took in hand to fly with wings, and to that effect caused make a pair of wings of feathers, which being fastened upon him, he flew off the castle wall of Stirling, but shortly he fell to the ground and brake his thighbone."

Nearly 250 years later the aeronaut had not made much progress, for we read of the Marquis de Bacqueville in 1742 attaching to his arms and legs planes of his own design and launching himself from his house in the attempt to fly across the Seine, into which, regrettably, he fell.

Meanwhile the seventeenth-century philosophers had been theorizing. In 1638 John Wilkins, the founder of the Royal Society, published a book entitled Daedalus, or Mechanical Motions. A few years later John Glanville wrote in Scepsis Scientifica "to them that come after us it may be as ordinary to buy a pair of wings to fly into remotest regions, as now a pair of boots to ride a journey," the sceptic proving a truer prophet than the enthusiast. By 1680 Giovanni Borelli had reached the conclusion, in his book De Volatu, that it was impossible that man should ever achieve flight by his own strength. Nor was he more likely to do so in the first aerial ship, designed in 1670 by Francesco Lana, which was to be buoyed up in the air by being suspended from four globes, made of thin copper sheeting, each of them about 25 feet in diameter. From these globes the air was to be exhausted, so that each, being lighter than the atmosphere, would support the weight of two or three men. A hundred years elapsed before Dr. Joseph Black of the University of Edinburgh made the first practical suggestion, that a balloon inflated with hydrogen would rise.

THE INVENTION OF THE BALLOON.

It was in 1783 that Montgolfier conceived the idea of utilizing the lifting power of hot air and invited the Assembly of Vivarais to watch an exhibition of his invention, when a balloon, 10 feet in circumference, rose to a height of 6,000 feet in under ten minutes. This was followed by a demonstration before Louis XVI at Versailles, when a balloon carrying a sheep, a cock, and a duck, rose 1,500 feet and descended safely. And on November 21st of the same year Pilatre de Rozier, accompanied by the Marquis d'Arlande, made the first human ascent, in the "Reveillon," travelling 5 miles over Paris in twenty minutes.

England, it is not surprising to learn, was behind with the invention, but on November 25th, 1783, Count Francesco Zambeccari sent up from Moorfields a small oilskin hydrogen balloon which fell at Petworth; and in August of 1784 James Tytler ascended at Edinburgh in a fire balloon, thus achieving the first ascent in Great Britain. In the same year Lunardi came to London and ballooning became the rage. It was an Englishman, Dr. Jefferies, who accompanied Blanchard in the first cross-Channel flight on January 7th, 1785. Fashionable society soon turned to pursuits other than watching balloon ascents, however, and the joys of the air were confined to a few adventurous spirits, such as Green and Holland, who first substituted coal gas for hydrogen and in 1836 made a voyage of 500 miles from Vauxhall Gardens to Weilburg in Nassau, and James Glaisher, who in the middle of the century began to make meteorological observations from balloons, claiming on one occasion, in 1862, to have reached the great height of 7 miles.

FIRST EXPERIMENTS IN GLIDERS AND AEROPLANES.

The world seemed content to have achieved the balloon, but there were a few men who realized that the air had not been conquered, and who believed that success could only be attained by the scientific study and practice of gliding. Prominent among these, Sir George Cayley, in 1809, published a paper on the Navigation of the Air, and forecasted the modern aeroplane, and the action of the air on wings. In 1848 Henson and Stringfellow, the latter being the inventive genius, designed and produced a small model aeroplane--the first power-driven machine which actually flew. It is now in the Smithsonian Institute at Washington. Of greater practical value were the gliding experiments by Otto Lilienthal, of Berlin, and Percy Pilcher, an Englishman, at the end of the last century. Both these men met their death in the cause of aviation. Another step forward was made by Laurence Hargrave, an Australian, who invented the box and soaring kite and eighteen machines which flew.

From the theoretical point of view, Professor Langley, an American, reached in his Experiments in Aerodynamics the important conclusion that weight could best be countered by speed. From theory Langley turned to practice and in 1896 designed a steam-driven machine which flew three-quarters of a mile without an operator. Seven years later, at the end of 1903, he produced a new machine fitted with a 52 horse-power engine weighing less than 5 lb. per horse-power; but this machine was severely damaged ten days before Wilbur Wright made his first flight in a controlled power-driven aeroplane.

THE WRIGHT BROTHERS AND THEIR SUCCESSORS IN EUROPE.

The Wright brothers directed their whole attention to aviation in 1899. By 1902, as the result of many experiments, they had invented a glider with a horizontal vane in front, a vertical vane behind, and a device for "warping" the wings. Their longest glide was 622-1/4 feet. This was followed by the construction of a machine weighing 600 lb., including the operator and an 8 horse-power engine, which on December 17th, 1903, realized the dreams of centuries.

After an increasing number of experiments, a machine built in 1905 flew 24-1/4 miles at a speed of 38 miles an hour. It is interesting to recall that the new invention was refused once by the United States and three times by the British Government.

It was not until September 13th, 1906, that Ellehammer, a Danish engineer, made the first free flight in Europe, his machine flying 42 metres at a height of a metre and a half. About the same time reports of the Wrights' successes began to reach Europe and were quickly appreciated by the French.

Space forbids that I should enter into the achievements of the early French aviators, among whom the names of Ferber, Bleriot and Farman will always rank high in the story of human faith, courage and determination. It is a record of rapid advance. Farman made a circuit flight of 1 kilometre in 1908, and flew from Chalons to Rheims, a distance of 27 kilometres, in twenty minutes. Bleriot crossed the Channel in a monoplane of his own design in forty minutes. French designers improved the control system, and French machines became famous. The records of the Rheims meeting of 1909 serve to illustrate the progress made during the first phase of aviation. Latham won the altitude prize by flying to a height of over 500 feet. Farman the prize for the flight of longest duration by remaining more than three hours in the air, and the passenger carrying prize by carrying two passengers round a 10-kilometre course in 10-1/2 minutes. The Gnome rotary engine was first used with success at this meeting.

Before turning to the pioneer efforts in England and the pre-war organization of our air forces, some account of the development of the lighter-than-air dirigible is desirable.

THE FIRST AIRSHIPS.

The earliest conception of an airship is to be found in General Meusnier's design in 1784 for an egg-shaped balloon driven by three screw propellers, worked, of course, by hand. The chief interest in his design, though it never materialized, lies in the fact that it provided for a double envelope and was the precursor of the ballonet. The first man-carrying airship was built by Henri Giffard in 1852. It had a capacity of 87,000 cu. feet, a length of 144 feet, a 3 horse-power engine, and a speed of 6 miles an hour. A gas engine was first used twenty years later in an Austrian dirigible, giving a speed of 3 miles an hour. About the same time much useful work was accomplished by Dupuy de Lome, whose dirigible, with a propeller driven by man power, gave a speed of 5-1/2 miles an hour. Twelve years later, in 1884, two French Army officers, Captain Kubs and Captain Renard, constructed the first successful power-driven lighter-than-air craft fitted with an 8-1/2 horse-power electric motor, which may be regarded as the progenitor of all subsequent non-rigid airships. In 1901 Santos Dumont flew round the Eiffel Tower, travelling 6-3/4 miles in 1-1/2 hours, and in 1903 the flight of the "Lebaudy," covering a distance of 40 miles at a speed of 20 miles an hour, led the French military authorities to take up the question of airships.

What the French initiated, the Germans, concentrating with characteristic thoroughness on the development of the rigid as opposed to the non-rigid airship, improved. In 1896 Wolfert's rigid airship attained a speed of 9 miles an hour and in 1900 the first Zeppelin was launched. Whatever we may think of the German methods of using their airships during the war, we cannot but admire the courage and determination of Count Zeppelin, who, in spite of many mishaps, succeeded in producing the finest airships in the world and inspiring the German people with a faith in the air which they have never lost. From 1905 onwards development was rapid. In 1907 Zeppelin voyaged in stages from Friedrichshaven to Frankfort, a distance of 200 miles in 7-1/2 hours. Popular enthusiasm is illustrated by the fact that within a few months the same airship made four hundred trips, carrying 8,551 passengers and covering 29,430 miles. Other airships showed similar records. Between 1909 and 1913 eighteen of the Parseval type were built, and 1912 saw the construction of the first Schutte-Lanz, designed expressly for naval and military purposes. If France at this period led the world in aeroplane design, Germany was undeniably ahead in airship development.

In Great Britain, in 1905, we had one very small airship, designed and constructed by Willows.

THE BEGINNINGS OF AVIATION IN ENGLAND.

Though the names of Pilcher, Dunne, Howard Wright, and Rolls testify to the fact that the science of aviation had its followers in England at the beginning of this century, flying came comparatively late, and the real interest of the movement centres round the early efforts of military aviation from 1912 onwards. Nevertheless this country could ill have dispensed with the experiments of that small and courageous band of aviators, among whom Dickson and Cody were prominent. By 1908 Cody had built an aeroplane and was making experimental flights at Aldershot. In 1907, A. V. Roe, working under great difficulties, constructed and flew his first machine, a triplane fitted with an 8-10 horse-power twin cylinder Jap bicycle engine, the first tractor type machine produced by any country, and a very important contribution to the science of flight. In 1910 and 1911 we find de Havilland, Frank Maclean and the Short Brothers, Ogilvie, Professor Huntingdon, Sopwith and the Bristol Company, starting on the design and construction of machines, of which the names have since become famous. At the same time certain centres of aviation came into existence, such as Brooklands, where I well remember beginning to fly in August, 1910, Hendon, Larkhill and Eastchurch, destined to be the centre of naval aviation. It is significant, however, of the slow progress made that by November 1st, 1910, only twenty-two pilot's certificates had been issued, and it was Conneau, a French naval officer, who in 1911 won the so-called "Circuit of Britain," i.e. a flight from Brooklands and back via Edinburgh, Glasgow, Exeter and Brighton. Cody and Valentine were the only British competitors to complete the full course.

In May 1911 a demonstration was organized by the owners of the Hendon Aerodrome to which a large number of Cabinet Ministers, members of parliament, and army and navy officers were invited. The War Office co-operated by arranging for a small force of horse, foot and guns to be secretly disposed in a specified area some miles distant and by detailing two officers, of whom I was one, to test what could be done to find and report them by air. I remember that I had a special map prepared, the first used in this, and I think any country, for the aeroplane reconnaissance of troops. After a sufficiently exciting trip, and with the troops successfully marked on the map, Hubert, my French pilot, and I, returned and made our report to General Murray, the Director of Military Training. It was a very interesting flight; the weather good; our height about 1,500 feet; the machine a 50 horse-power Gnome "box-kite" Henri Farman, which at one period of our 35 mile an hour return journey elected to point itself skywards for an unpleasant second or two and fly "cabr?; I can see Hubert now anxiously forcing his front elevator downwards and shouting to me to lean forward in order to help to bring the nose to a more comfortable bearing!

Many pages could be filled with the difficulties and exploits of the first British aviators, but enough has been said to show that, compared with that of aeroplanes in France and of airships in Germany, development in this country started late, progressed slowly and excited little public interest. The work of the pioneers was, however, not in vain, since it opened the eyes of our military authorities to the value of aviation and led to the formation of that small but highly efficient flying corps which during the war expanded into an organization without rival. Let us now turn to the inception of the air forces of the Crown and the position with regard to these and to air tactics at the outbreak of war.

THE INCEPTION AND DEVELOPMENT OF AIRCRAFT AS PART OF THE FORCES OF THE CROWN.

Nations have tended to regard flight as a prerogative of war. A balloon school was formed in the early days of the French revolutionary wars; the French victory at Fleurus in 1794 was ascribed to balloon reconnaissance; balloons were used by the Federal Army in the American Civil War, and during the Siege of Paris Gambetta effected his escape by balloon in 1871.

The Balloon Factory.

In England experiments were begun at Woolwich Arsenal in 1878, and in 1883 a Balloon Factory, a Depot and a School of Instruction were established at Chatham. The expedition to Bechuanaland in 1884, under the command of Sir Charles Warren, was accompanied by a detachment of three balloons, and the following year balloons co-operated with the Sudan Expeditionary Force, when Major Elsdale carried out some photographic experiments from the air.

In 1890 a balloon section was introduced into the Army as a unit of the

Royal Engineers, and not long afterwards, the Balloon Factory was established at South Farnborough, where in 1912 it was transformed into the Royal Aircraft Factory. Four balloon sections took part in the South African War and were used during the Siege of Ladysmith, at Magersfontein and Paardeburg. Colonel Lynch, who served in the Boer Army, stated at a lecture delivered in Paris after the war that "the Boers took a dislike to balloons. All other instruments of war were at their command; they had artillery superior for the most part to, and better served than, that of the English; they had telegraphic and heliographic apparatus; but the balloons were the symbol of a scientific superiority of the English which seriously disquieted them."

I went through a course in ballooning during leave from West Africa in 1904 and remember that partly owing to the energy of Colonel Capper, partly to the impetus given by the South African War, and partly to the growing interest in all things aeronautical throughout the civilized world, it was noticeable that the activities of the Balloon Factory were increasing in many directions. Although the spherical balloon had been improved, its disabilities were recognized and experiments were made with elongated balloons, man-flying kites, air photography, signalling devices, observation of artillery fire, mechanical apparatus for hauling down balloons, and petrol motors. A grant for a dirigible balloon was obtained in 1903, though it was not until 1907, the year in which Cody began the construction of his aeroplane at Farnborough, and Charles Rolls his experiments, that the airship "Nulli Secundus" made her first flight. She was about 120 feet long and 30 feet in diameter, and was driven by a 40 horse-power engine at a speed of 30 miles an hour. On October 5th this airship flew to London in an hour and a half, circled round St. Paul's, man[oe]uvred over Buckingham Palace, and descended at the Crystal Palace. In the same year, be it remembered, a Zeppelin had made a trip of 200 miles from Friedrichshaven to Frankfort. The "Nulli Secundus" was followed in 1910 by the "Beta" and the "Gamma."

Meanwhile an Advisory Committee for Aeronautics had been appointed, and the National Physical Laboratory had organized a department at Teddington for the investigation of aeronautical problems in co-operation with the Balloon Factory.

The Air Battalion.

In 1911 the authorities could no longer close their eyes, especially at a time when rumours of war were rife, to the rapid development of heavier-than-air craft on the Continent. So far, as we have seen, the aeroplane had been regarded in England as little more than the plaything of a few adventurous but foolhardy spirits. A certain amount of experience in piloting and handling aeroplanes had been gained by a handful of Army officers, but the machines used either belonged to the officers themselves, to civilians, or to aviation firms. I was at that time a general staff officer in the Directorate of Military Operations under General Wilson, now Field Marshal and late Chief of the Imperial General Staff, and was the only officer in the War Office who had learned to fly. It appeared very important that a study of the military possibilities of aviation should be made. The prime role of cavalry, reconnaissance, seemed to have passed from it. In addition to my normal duties, I visited France, Germany and Italy, collected information on foreign activities, wrote reports, and tried to create a knowledge of the possible effect of future military aeronautics and to urge the formation of a flying corps.

In 1911 the Air Battalion of the Royal Engineers, consisting of Headquarters, No. 1 Company (Airships) and No. 2 Company (Aeroplanes), was formed and superseded the Balloon School. The creation of No. 2 Company, stationed at Larkhill, marked the first formation of a British military unit composed entirely of heavier-than-air aircraft. The same year witnessed the inception of the B.E., F.E. and S.E. type machines in the Balloon Factory, but the total of our machines, both for naval and military requirements, amounted to something less than twelve aeroplanes and two small airships; and the mishaps suffered by the military machines on their flight from Larkhill to Cambridge, to take part in Army Man[oe]uvres, were significant of their unreliability.

The Royal Flying Corps.

In view, therefore, of the reports received of the progress abroad, the Air Battalion was clearly insufficient to meet the demands which might be made upon it in the event of war; and at the end of 1911 the Prime Minister instructed a standing Sub-Committee of the Committee of Imperial Defence to consider the future development of air navigation for naval and military purposes. As a result of their deliberations the Committee recommended the creation of a British Air Service to be regarded as one and designated the Royal Flying Corps; the division of the Corps into a Naval Wing, a Military Wing, and a Central Flying School; the maintenance of the closest possible collaboration between the Corps, the Advisory Committee for Aeronautics and the Aircraft (late Balloon) Factory; and the appointment of a permanent Consultative Committee, named the Air Committee, to deal with all aeronautical questions affecting both the Admiralty and the War Office.

Consequent upon these recommendations, a Technical Sub-Committee was formed, consisting of Brig.-General Henderson, Major MacInnes of the directorate of Military Training at the War Office, a splendid officer, who died during the war, and myself, to draft the new scheme. The objects kept in view in framing our peace organization were to suit it to war conditions, as far as they could be foreseen, to base it on an efficient self-contained unit, and, while allowing for the wide differences between naval and military requirements, to ensure the maximum co-operation between the two branches of the Service. Success beyond expectation was achieved in the first two objects, but, as will be seen, the naval and military branches tended for unforeseen but good reasons to diverge, until they joined hands again in 1918 as the Royal Air Force. The bases of the military organization were, a headquarters, the squadron, and the flying depot. These proved their value during the war and have remained the units of our air forces to this day. The Military Wing was to form a single and complete organization and contain a headquarters, seven aeroplane squadrons, each to consist of twelve active machines and six in reserve, one airship and kite squadron, and a flying depot. All pilots, whether of the Naval or the Military Wing, were eventually to graduate at the Central Flying School, whence they could join either the Naval Wing at Eastchurch or one of the Military Squadrons. In time of war each branch of the Service was to form a reserve for the other if required.

The Military Wing.

In accordance with this scheme I received instructions to organize, recruit, train and command the Military Wing of the Royal Flying Corps. The functions of the Military Wing were quite clear: it was to meet the air requirements of the Expeditionary Force primarily for reconnaissance purposes, but its organization was framed so that it could easily be expanded and the scope of its duties widened. Headquarters were established at Farnborough on May 13th, 1912: Barrington-Kennett, an officer of the Grenadier Guards who had been attached to the Air Battalion, was appointed, and made the best of all possible adjutants; and the nucleus of the Corps, consisting at first of the cadres of an airship squadron under Edward Maitland, of two aeroplane squadrons under Burke and Brooke-Popham, and a flying depot (later the aircraft park) under Carden, who was a little later greatly assisted in the complex matter of technical stores by Beatty, came into existence. At the same time the construction of the Central Flying School was started at Upavon, under Captain G. Payne, R.N. With regard to the other squadrons provided for, the nucleus of No. 4 Squadron was formed the same year, and that of No. 5 Squadron the following year, of Nos. 6 and 7 Squadrons in 1914, while No. 8 Squadron was not started until after the outbreak of war.

Records of the progress and growth of the Corps were left at Farnborough when the Headquarters and four squadrons went to France in August, 1914, and have been lost. This is particularly unfortunate because without them it will be difficult for the historian of the Corps adequately to describe the beginnings and to assess the value of the work then carried out.

The task of forming the new service, which was to do much to assist the Army in saving England, was begun. The time was very short. A great energy had to be brought to the work. As with all things new, it had to contend with apathy and opposition on all sides. There was no precedent to help. The organization of the Corps to its smallest detail of technical stores, supply and transport had to be thought out. The type of machine required; the method of obtaining it from a struggling industry; its use and maintenance; the personnel, its training and equipment; these, and a thousand other aspects of the question, required the employment of a large staff of experts. But the experts did not exist and the duties were carried out almost entirely at Farnborough, where in addition time had to be found to compile the official training and other text books and regulations required for an entirely new arm.

In addition to the innumerable problems inherent in the organization, growth and training of the Military Wing, the two years between its inception and the outbreak of war were strenuously applied to solving the problems of air tactics and strategy. Until the South African War the British Army had been drilled under the influence of stereotyped Prussian ideas. Perhaps the South African War led too far in an opposite direction, but it taught us one thing, which was to prove of such importance in 1914--the value of mobility; and we realized in aircraft the advent of the most mobile arm the world has yet seen.

All was new. A new Corps. A new element in which to work. New conditions in peace akin to those in war. And there had to be developed a new spirit, combining the discipline of the old Army, the technical skill of the Navy, and the initiative, energy and dash inseparable from flying. There were the inevitable accidents, but training had to be done. We existed for war and war alone would show whether we had thought and worked without respite aright. We had to prove our value to the other arms, many of the leaders of which, owing to a long period of peace, found difficulty in differentiating between the normal usages of peace and war and in understanding the right use of aircraft. Somehow or other time was found during 1912, 1913 and 1914 to write to reviews, to lecture at army and other centres of training, to attend Staff rides, and to endeavour in every way possible to learn how best to work in with the army commands and to teach those commands the usefulness and limitations of aircraft.

As Ruskin wrote:

"Man is the engine whose motive power is the soul and the largest quantity of work will not be done by this curious engine for pay, or under pressure, or by the help of any kind of fuel which may be supplied by the cauldron. It will be done only when the will or spirit of the creature is brought to its own greatest strength by its own proper fuel, namely the affection."

I was intensely proud of my command and often thought of the time when, as I had been promised, I should, in the event of war, command it in the field. We worked at white heat believing that war was coming soon; believing that our efforts would have a real effect on the result; and determined that the new arm should rank second to none among the forces of the Crown. Esprit de Corps was of vital importance, but as officers and non-commissioned officers were drawn from every branch and every regiment of the army this was no easy matter and was only achieved by the splendid example and precept of such men as Herbert, Becke, Longcroft, Chinnery and Barrington-Kennett. We selected our motto: "Per Ardua ad Astra." It was in this atmosphere that the Military Wing grew in peace. It was in this atmosphere that the soul was formed which later under the great strain of war impelled our pilots forward cheerfully to face every duty and every danger in the true spirit of manliness and fearless confidence.

As in framing the original scheme on paper, so in giving it life it was our aim to organize the Corps, so that, whatever its future strength, it would be sound and efficient, and its continuity of growth effected without even temporary dislocation or waste. The tactical unit of the Military Wing--the squadron, consisting of three flights, each of four machines with two in reserve--had the advantage that it was of sufficient size to act independently, while it was not too unwieldy for a single command. It was equally suitable for independent or co-operative action, and the full complement of seven squadrons would, in addition to a reserve, furnish one squadron for each division of an Army

Expeditionary force of the size then contemplated, though no definite allotment of aeroplanes to the lower commands was at first intended. The French and Germans, on the other hand, were building up their organizations with smaller units, with the result that they found even greater difficulties than ourselves in obtaining sufficient experienced officers to command them. It is probable that the consequent lack of concentration, knowledge and determination to stick to sound principles of action was one of the causes underlying the non-success of the German air service in the opening phases of the war.

According to the system employed squadrons were formed, organized, equipped, and a certain amount of preliminary training carried out, at Farnborough, when on completion the squadron moved to one of the stations which I had established or was forming at Netheravon, Montrose, Gosport, Dover, and Orfordness, Netheravon being the largest. This dispersion of squadrons did not affect the entity and cohesion, under Wing headquarters at Farnborough, of the Corps as a whole. No. 3 Squadron, one of the original two referred to, removed to Netheravon from Larkhill in June.

Similarly, and in order to avoid congestion at Farnborough, to foster a spirit of self-support and to enable air operations to be carried out with troops in Scotland, No. 2 Squadron was sent to Montrose. Five of its machines flew all the way, and it became one of the principles of training that machines should fly whenever a move was ordered. Thus in 1913 six machines from this squadron were flown from Montrose to Limerick--a great feat then--to take part in the Irish Command man[oe]uvres, the crossing of the Irish Channel being successfully carried out both ways by all machines. Another flight of an experimental nature was made by Longcroft, with myself as passenger, from Farnborough to Montrose in a single day with only one landing.

The unavoidable and never-relaxing strain inherent in the daily and hourly use of an instrument, in the design, maintenance and improvement of which we could only grope our way, was very great. In peace before the war, as later in the war, the only variation to strain lay in periods of increased strain.

At Headquarters, in addition to the normal duties of command and co-ordination, and the supply of all technical stores to squadrons, there was carried out all recruiting, and I also formed a specialized flight for the study of technical problems, such as the use of wireless from aircraft. The bulk of experimental work was originally undertaken by the Royal Aircraft Factory, under the Superintendent, Mr. O'Gorman, who always helped us in every way possible, but by 1913 I felt it necessary to enlarge the duties of the special flight and an Experimental Section was formed at Wing Headquarters at Farnborough with an officer, Musgrave, in charge. In addition, for each squadron an officer was appointed Squadron Officer for Experiments, thus ensuring the diffusion of information throughout the Corps, and affording the opportunity to each unit of carrying out the experiments best suited to the material and apparatus at its command. Similarly other individual officers were detailed in each squadron on a co-ordinated scheme, for such duties as Officer-in-charge of Stores, Workshops, Mechanical Transport, Meteorology, etc.

The formation at Farnborough of the Line of Communications R.F.C. Workshop or Flying Depot--later known as the Aircraft Park--completed the organization of the Military Wing.

I was very anxious as early as possible to prove the structure as a unified self-supporting, mobile and easily handled flying corps as far as it had gone, and in June, 1914, this was done by the concentration in camp at Netheravon of the entire Military Wing, comprising Headquarters and Headquarters Flight, the four completed squadrons and the nucleus of No. 6 Squadron, the Aircraft Park and a detachment of the Kite Section. Mobilization, a very difficult process when it came, would have been almost impossible had the concentration not taken place. The object of the camp was a month's combined training to test personnel, both in the air and on the ground, and the handling of aircraft and transport both by day and night. Endeavours were made to solve by means of lectures, discussions and committees the problems connected with mobilization, technical and military training, observation, wireless telegraphy, signals, night flying, photography, bomb-dropping, workshops, stores, meteorology, transport, shifting of camp and aerodrome, supply and maintenance of units in the field, etc.--in fact the whole organization essential to the efficiency and cohesion of a Flying Corps, under conditions as similar as possible to those expected on active service. Very valuable experience was obtained from the work carried out. The necessarily wide gaps in our knowledge were brought home in more concrete form. It was also evident that the force was very small. But within three months it was proved under the strain of war that the organization and training had been laid down on sound principles.

The Naval Wing.

As in the case of the Army, it was to airships that the Navy first turned its attention, and the birth of naval aviation may be said to date from July 21st, 1908, when Admiral Bacon submitted proposals for the construction of a rigid airship, the ill-fated "Mayfly" which was destroyed on her preliminary trials. The Admiralty thereupon decided to discontinue the construction of airships, the development of which was left to the Army until May, 1914, when it was decided that all airships--that is No. 1 Squadron of the Military Wing--should be taken over by the Naval Wing. This was partly the result of a report by two Naval Officers, who visited France, Austria and Germany, as was the purchase of two vessels of the Parseval and Astra Torres types, and a small non-rigid from Willows. The construction of a number of other airships was ordered, but for various reasons was delayed or never completed up to the outbreak of war.

Although at first sight the functions of the Naval Wing--coast defence and work with the Fleet--seemed hardly more difficult to perform than those of the Military Wing, in practice, as I was to find later from personal experience when in command of the R.N.A.S. at Gallipoli, they were more complicated, while the slowness of the Admiralty in evolving a clear scheme of employment and a definite objective made itself felt. Before the war the achievements of the Naval Wing were due rather to individual effort than to a definite policy of organized expansion. It was the pilot and the machine rather than the organization which developed.

As already stated, Eastchurch was chosen by the Short Brothers for their experiments in aeroplanes in 1909, but it was not until 1911 that the Admiralty bought two machines and established the first Naval Flying School at that place. The same year Commander O. Swann purchased from Messrs. A. V. Roe a 35 horse-power biplane and began to carry out experiments with different types of floats, as a result of which a twin-float seaplane was produced--the first to rise off the water in this country.

For some time seaplanes were in a very experimental stage and at best could only rise from, and alight on, calm water, though it is interesting to note that as far back as 1911 the employment of seaplanes for torpedo attack, which I think will be one of the most important developments of aircraft in the future, engaged the attention of the Navy, and a Sopwith seaplane carrying a 14-inch torpedo made its first flight at Calshot in 1913. For this reason, therefore, it appeared that principally aeroplanes and airships would have to be employed from shore bases for coast defence and that "carrier" ships would be necessary to enable seaplanes to work with the Fleet.

The first stations set up were Eastchurch, Isle of Grain, Calshot, Felixstowe, Yarmouth, Cromarty and Kingsnorth, from which at the outbreak of war an organized coastal patrol was established.

From the outset the Naval Wing, assisted by its large percentage of skilled technical personnel, paid great attention to experimental work of all sorts. Thus in 1912 the detection of submarines by aircraft was taken up, in 1913 valuable results were obtained from bomb-dropping, and a large number of experiments in wireless, machine gunnery and fighting carried out. In addition, efforts were made to extend the power, range and capacity of engine and machine.

The second Naval problem, that of co-operation with the Fleet, involved the flight of aircraft from ships and the design of aircraft carriers. In 1911 an aeroplane for the first time took off successfully from the deck of a cruiser at anchor, and the following year an aeroplane flew from H.M.S. "Hibernia," while under weigh, but it was not until after the outbreak of war that alighting on decks was successfully accomplished. The first ship to be fitted up as a parent ship for seaplanes was the "Hermes" in 1913.

These specialized technical requirements and developments explain why the Naval Wing and the Royal Naval Air Service tended towards individualism rather than cohesion. While the Military Wing, or Royal Flying Corps, progressed further as an organized fighting force, the Royal Naval Air Service, amongst the 100 odd aeroplanes and seaplanes on charge which were mainly of the Short, Sopwith, Avro, Farman and Wright types, possessed in 1914 the more powerful engines and a number of aeroplanes fitted with wireless and machine guns, while their bomb-dropping arrangements were also in a more advanced stage of development.

An Air Department was formed at the Admiralty in 1912 to deal with all questions relating to naval aircraft. Naval officers were trained from the beginning at Eastchurch rather than at the Central Flying School, and in 1913 the appointment of an Inspecting Captain for Aircraft, with a Central Air Office at Sheerness as his headquarters, accentuated a growing tendency for the Naval Wing to work on independent lines.

The Naval Wing grew rapidly and in the middle of 1914 was reorganized as the Royal Naval Air Service, comprising the Air Department of the Admiralty, the Central Air Office, the Royal Naval Flying School, the Royal Naval Air Stations, and all aircraft, seaplane ships and balloons employed for naval purposes. This placed the naval air force on a self-supporting basis and the entity of the Royal Flying Corps as a whole, as originally provided for, was lost.

TACTICS AND THE MACHINE.

The value of the application of flying to war requires little demonstration.

The most important attributes of generalship are quick appreciation of a situation and quick decision. To the ordinary Commander the absence of information is paralysing. In the nineteenth century the mass of cavalry was the special instrument of information and to obtain it contact with the enemy's main forces had to be effected. It thus acted as a shield and also tried to provide the information necessary to enable the infantry to take the offensive.

Aviation, by the wide field of observation it commands, by the speed with which it can collect and transmit information, to a great extent lifts the fog of war and enables a general to act on knowledge where before he acted largely on deduction. Information once obtained, its mobile and far-reaching offensive power introduces the element of surprise, and permits of lightning strokes against the enemy's vital points.

Before the war reconnaissance was regarded as the principal duty of the aircraft of the Military Wing. This was due to two reasons, first, the obvious one that aircraft possessed advantages shared by no other arm for obtaining information quickly and over wide areas and reporting to Headquarters, and second, that experiment had proved the difficulty of loading aeroplanes with offensive weapons, such as bombs or machine guns, without impairing speed and climb.

The following statement, which I drafted and which was issued by the General Staff before the Army Man[oe]uvres of 1912, summarizes the position:--

"As regards strategical reconnaissance," it says, "a General is probably now justified in requiring a well-trained flyer, flying a modern aeroplane, to reconnoitre some 70 miles out and return 70 miles. This would be done at a speed of, say, 60 miles an hour in ordinary weather over ordinary country. Thus within four hours, allowing a wide margin, a report as to the approximate strength, formation and direction of movement of the enemy, if he is within a 70-mile radius, should be in the hands of the Commander."

To those imbued with a knowledge of military history this new method of ascertaining the enemy's movements might well seem revolutionary.

Let us take two instances illustrating what aircraft, with a radius of little over 100 miles, might have done in previous campaigns. For the operations which terminated in the capitulation of Ulm in 1805 Napoleon concentrated two army corps at Waltzburg and five along the left bank of the Rhine between Mannheim and Strasburg, his main body of cavalry under Murat being at the latter place. The Austrian Army under Mack was behind the Iller between Ulm and Memmingen, and expected the French to advance through the defiles of the Black Forest, where Napoleon did actually make a feint with his cavalry. Napoleon, however, crossing the Rhine on September 26th, 1805, moved east, and it was not until October 2nd, when the French Army had reached the line Ansbach, Langenburg, Hall and Ludwigsburg, and his envelopment was far advanced, that Mack realized that the main French advance was coming from the north. Aeroplanes of the type we possessed in 1914 could have reconnoitred the whole of Napoleon's preliminary position, could have detected his line of advance, especially as it was concentrated on a very narrow front, and could have brought back the information to the Austrian Headquarters within a few hours.

Aircraft would have been of even greater value on August 16th, 1870, at the Battle of Rezonville, where neither the French nor the Germans were aware of the other's movements. On the 14th a battle had been fought east of Metz which had resulted in the French retreat. On the morning of the 16th Moltke thought the French had retired west by the Metz-Verdun road and those to the north of it, and consequently he directed his left wing due west towards the Meuse to head off the French, sending his right army towards Rezonville to harass their rearguard. The French retreat, however, had been slow and two corps were still at Rezonville, while three corps and the reserve cavalry were within easy reach, some 130,000 men in all. At 9 in the morning the German 3rd Corps, unaided and far from support, attacked a position within reach of the whole French Army, believing it had to deal with a rearguard only. Bazaine, on the other hand, thinking that he was faced by the German main army, remained on the defensive, and lost the opportunity of defeating in detail first the 3rd and then the 10th German Corps. A few aeroplanes operating on a radius of 30 miles would have disclosed between daybreak and 10 a.m. the true position to either commander. Neither the German nor the French cavalry, though both were engaged, obtained any reliable information.

The problem as to how far aircraft would reduce the value of cavalry was widely discussed before the war. It was seen that by day aircraft could obtain quicker and more accurate information, but that cavalry retained their power of night reconnaissance, of mobile offensive action and of pinning the enemy to his ground by fighting. This was found to be so during the retreat, when, in addition to the direct value of aircraft for long-distance reconnaissance, an indirect asset of great importance lay in the release of the cavalry for battle action in assistance of the infantry. The question has become more acute since the offensive action of aircraft against ground targets has developed, but although we must never forget the splendid work of the mounted arm during the Retreat from Mons, and in March, 1918, factors have arisen tending to make the use of cavalry a problem of extreme difficulty in European wars, and it is possible that, in addition to their reconnaissance functions, aircraft will supersede the shock tactics and delaying action of cavalry, though this may be modified if, the sabre being a thing of the past, cavalry are converted into mounted machine gunners.

Air tactics and training were, therefore, chiefly studied from the point of view of reconnaissance. In addition to the possibility of being shot at by other aircraft, an important consideration was vulnerability from the ground. Before the war reconnaissances were carried out at heights varying from 2,000 to 6,000 feet, but it was generally considered that the aeroplane was safe from fire from the ground at heights above 3,000 feet.

Serious difficulties affecting the mobility of aircraft were the means of providing a regular supply of fuel and the selection of landing grounds when moving camp, which had to be close enough behind the front line as not to entail waste of time in flying out and back over friendly territory. This was later brought home to us in a very acute form during the Retreat from Mons.

As machines improved, increasing attention was paid to bettering their power of reconnaissance by air photography, their value in co-operation with artillery by wireless equipment, their offensive action by bomb dropping and their offence and defence by armament.

The value of a correct initiative and the aeroplane's role as an offensive weapon were fully appreciated and brought out in the Training Manual of the Royal Flying Corps which we compiled at Farnborough, and which was published early in 1914 by the War Office. It says:--

"It is probable that one phase of the struggle for the command of the air will resolve itself into a series of combats between individual aeroplanes, or pairs of aeroplanes. If the pilots of one side can succeed in obtaining victory in a succession of such combats, they will establish a moral ascendancy over the surviving pilots of the enemy, and be left free to carry out their duties of reconnaissance. The actual tactics must depend on the types of the aeroplanes engaged, the object of the pilot being to obtain for his passenger the free use of his own weapon while denying to the enemy the use of his. To disable the pilot of the opposing aeroplane will be the first object. In the case of fast reconnaissance aeroplanes it will often be advisable to avoid fighting, in order to carry out a mission or to deliver information; but it must be borne in mind that this will be sometimes impossible, and that, as in every other class of fighting, a fixed determination to attack and win will be the surest road to victory."

Speaking generally, the evolution of the machine, as apart from the engine, which hung behind, followed upon the evolution of air tactics. As soon as experience, often hard won at the cost of a valuable life, opened up new fields of activity for aircraft, the designer and constructor evolved new designs to meet the new requirements. It was no small achievement in this period to have solved the problem of inherent stability, both in theory and practice, so successfully, that from the aerodynamic standpoint our machines in 1914 compare favourably with those in use at the end of the war.

In dealing with the evolution of the machine during the three years prior to the war there are three landmarks: in the autumn of 1911 the few machines belonging to the Air Battalion failed to reach their destination for Army Man[oe]uvres; in May, 1912, the Royal Flying Corps was formed and experiments with a view to meeting military requirements were for the first time energetically and methodically prosecuted; and in August, 1914, four squadrons flew to France with machines which had attained a high degree of stability and were not inferior to any of those possessed by other countries. When it is remembered in what a short time these machines were evolved, it is not surprising that attention had been chiefly confined to the problem of the 'plane and stability, the engine and speed and reliability. Wireless, bombing, photography, night flying and machine gunnery had been discussed and experimented with, but no progress was made comparable to that effected under war conditions.

Machines and engines before the war were chiefly French. It is interesting to note those with which No. 3 Squadron, one of the first to be formed, commenced its career in May, 1912. They consisted of one 50 horse-power Gnome Nieuport, one Deperdussin, which by the way was privately owned, one Gnome Bristol, two Gnome Bleriot monoplanes, one Avro and one Bristol box-kite biplane. By September, 1912, the Squadron possessed fourteen monoplanes, but in that month, owing to the number of accidents incurred by them, the use of monoplanes was temporarily forbidden, and it was not until April, 1913, that the Squadron was fully equipped with B.E. and Maurice Farman biplanes organized in flights.

These types formed the backbone of the Military Wing, which also included Codys, Breguets, Avros, and, later, Sopwiths. The B.E.2c was produced by the Royal Aircraft Factory in the autumn of 1913 and demonstrated its high degree of stability by flying from Aldershot to Froyle and from Froyle to Fleet, distances of 6-3/4 and 8 miles respectively, without the use of ailerons or elevators. The progress made is illustrated by the fact that at the Army Man[oe]uvres of 1913 twelve machines covered 4,545 miles on reconnaissance and 3,210 miles on other flights, accurate observations being made from a height of 6,000 feet, without serious mishap.

In 1913 I recommended the gradual substitution of B.E.'s for Farmans on the ground of the all-round efficiency and superior fighting qualities of the former, and to secure the advantage of standardization, but it was objected by the War Office that the Farmans were the only machines that could mount weapons in front--an objection which was not met until firing through the airscrew was introduced--and that the slower Farmans offered greater advantages for observation, an idea which was long prevalent. As a result, a compromise was effected, and two squadrons were equipped with B.E.'s and two with homogeneous flights of Farmans, Bleriots and Avros.

At the outbreak of war the most successful machines possessed by the Military Wing were the B.E.2 tractor with a 70 horse-power Renault engine, a speed of 73 miles an hour, and a climb of 3,000 feet in nine minutes; and a Henri Farman pusher with a speed of 60 miles an hour, and a climb of 3,000 feet in fourteen minutes. A special study was being made in 1914 of the best methods of ensuring clear observation of the ground, and partly in this connection staggered planes were introduced, culminating in the B.E.2c's, which were not, however, available for service in any numbers until 1915.

To sum up, the technical development of aircraft has taken place, and will continue side by side with the evolution of the uses to which aircraft can be put. While due attention was paid to problems connected with the anticipated duties of aircraft ancillary to that of reconnaissance, owing to the short space of time between the formation of the Royal Flying Corps and the outbreak of war, to the difficulties connected with the engine, and to causes inseparable from peace conditions, development had been more or less confined to evolving a stable and reliable machine with a good field of view.

CONCLUSIONS.

The foregoing outline of the development of aviation from the earliest times up to the war--a story of human endeavour and achievement in the air with its attendant dangers and difficulties--is not without value in endeavouring to assess that which has since occurred.

At the beginning of the year 1912 the Royal Flying Corps did not exist. At the beginning of the Great War, in 1914, England found herself with an air service which, though much smaller than those of Germany or France, was so excellently manned and organized, trained and equipped, that it placed her at a bound in the front rank of aviation.

The machine was stable, but the engine still unequal to the tasks laid upon it. Civil Aviation practically did not exist.

I shall now describe the extension of air duties under war conditions; the increasing value of aircraft for general action and air tactics and their development and far-reaching effect as the right hand of strategy. This resulted in the expansion of our flying corps from a total of 1,844 officers and men, and seven squadrons with some 150 machines fit for war use, to a total of nearly 300,000 officers and men, and 201 squadrons and 22,000 machines in use at the end of the war, and in the evolution of the machine to a point where we can regard it, not only as a weapon of war, but as a new method of transport for commercial purposes in peace.

CHAPTER II

WAR

GENERAL REMARKS ON WAR DEVELOPMENT.

In dealing with the story of the beginnings of aviation and the evolution of aircraft up to the war, we have seen that though its growth was infinitesimal compared with that which came with the impetus of war, the air service took definite and practical shape more rapidly than had up to that time any other arm of the Army or Navy in peace.

In 1914 we had reached a point where we possessed a small but mobile and efficient flying force, equipped and trained essentially for reconnaissance. Although experiments had been made, little had been achieved in the use of wireless from aircraft, air photography, bomb dropping, armament or the development of air fighting. As with the Army and Navy, war quickened and expanded all the attributes of air operations in a way which could not have been foreseen before the struggle occurred; and, as it would have been impossible for the Army and Navy to build up their war organization without the foundation of the pre-war service, so it was the splendid quality of the original Royal Flying Corps that made this expansion possible.

Before the war the Royal Flying Corps was considerably smaller than the air services of either France or Germany, and to attain even the strength with which the Military Wing left England the bulk of the trained officers and men, and almost all the machines fit for service, had to be taken. When I started to raise the Corps, in May, 1912, the War Office estimated that its organization, (of a headquarters and seven aeroplane and one airship squadrons) would take at least four years; instead, there had been little more than two. Even at the risk of leaving insufficient personnel and material behind to form and train new squadrons, I recommended that four complete squadrons (including the wireless machines which had to be thrown in to make up the numbers) should be sent overseas to help the British Expeditionary Force in bearing the brunt of the terrific blow that was to come. It was a very serious matter that so little could be left with which to carry on in England, but we considered it essential to dispatch at once to France every available machine and pilot, because both political and military authorities were of opinion that for economic and financial reasons a war with a great European power could not last more than a few months. Another reason was that those of us who had been at the Staff College during the few years before the war, or who had recently served on the General Staff at the War Office, believed that the weight of the German attack would be made through Belgium, where, owing to the enclosed nature of the country, cavalry would be at a disadvantage, and we realized therefore, and urged, the great effect which the air would have from the commencement of operations--a view which was not widely held, especially among senior officers in the Army. We also felt the necessity of using our maximum air strength from the outset, so as to prove its supreme importance as quickly and practically as possible. It required the Retreat from Mons before even G.H.Q. as a whole would accept the fact, though Colonel Macdonogh, the head of the intelligence section, was our firm ally. The iron of confidence, both to used and user, had to be welded with the first great blows on the anvil of war. For these reasons it was vital that every available trained pilot and suitable machine should be employed with the Army, even at the danger of serious initial depletion at Home. The smooth progress of expansion was largely attributable alike to the strength of the pre-war spirit, organization and training,[2] and to the results actual and moral obtained by the first four squadrons during the Retreat and the following weeks of the war under centralized control. The French distributed their "Escadrilles," which were approximately of the size of our "flight," from the beginning, and it is probable that one cause of failure in the German air service during the same period lay in the initial dispersion of units and lack of unified control by the higher command. The British Expeditionary Force having been saved during the Retreat, Paris having been saved at the Marne, the great German army having made a retirement, a lengthy war of position having become obvious, confidence in the air service, both within and without, having been established, the centralized system necessarily adopted up to that time could be relaxed, and we were able to send home officers and men with greatly increased experience to help build up the many new squadrons which would be required to co-operate with the new armies.

[2] On October 17, 1914, Sir J. French wrote: "Such efficiency as the R.F.C. may have shown in the field is, in my opinion, principally due to organization and training."

Gradually, as the numbers in the field permitted, increased duties were undertaken. The Army, though it did not do so at first, yet came to understand the immense importance to itself of air reconnaissance. So much so indeed that our machines and pilots were generally many too few to attempt more than the absolute essentials, and calls were often made upon them which were beyond their strength to meet. An ironic contrast to this was supplied, however, at the evacuation of the Dardanelles, where I was commanding the air service (the R.N.A.S.), and was asked to be careful not to do too much air work. This at a time when through stress and strain and loss we had, I think, a total of five machines left able to take the air!

Observation was, and remains, the prime purpose for which the Royal Flying Corps was formed. 1914 was a year of reconnaissance, but with the advent of trench warfare at the Battle of the Aisne, the first attempts were made to extend its scope by the use of wireless for artillery co-operation, and by air photography, both of which developed rapidly. Headway was also being made with bombing. Then machines carrying out their special duties had to be protected, while it became necessary to prevent hostile machines from effecting similar functions, with the result that 1915 saw the beginnings of systematic air fighting.

In 1915 the easily man[oe]uvrable Fokker, with its machine-gun synchronizing gear for firing through the propeller, gave the Germans a temporary lead, but by the Battle of the Somme this was outclassed and in 1916 our air superiority became marked. The Royal Flying Corps was by that time organized into Brigades and Wings, one Wing operating with each Army for fighting and distant reconnaissance, and one Wing with each Corps for short reconnaissance and such specialized work as artillery co-operation and contact patrols. Both types of machine took part in bombing operations.

There is generally perhaps a tendency, when reviewing the army and air effort in the war, to deal almost entirely with the Western Front and to forget the prodigious work done in many other theatres.

In 1915 the Royal Naval Air Service carried out all air work with the Army and Navy in the Gallipoli campaign and showed how a single air force could effect really important co-operation with both services. In addition to the normal duties of co-operation with the Army and the Fleet, and in spite of the difficulties of transport, supply and workshop arrangements, photographs were taken from the air of the greater part of the Peninsula, and the original inaccurate maps corrected therefrom; frequent bombing raids were carried out against objectives on the Peninsula, the Turkish lines of communications, and even Constantinople itself. In this campaign, too, torpedoes were used for the first time by aircraft and three ships were destroyed in the Dardanelles by this means. The distance from the hub of affairs, a line of supply about 6,000 miles in length, sickness and the climatic and geographical conditions rendered maintenance very difficult. Sand and dust driven in clouds by high winds greatly shortened the working life of engines. The heat during the summer caused the rapid deterioration of machines, while long oversea flights entailed loss from forced landings. There are many aspects of the deepest interest to be brought out when a complete history of the Campaign in Gallipoli comes to be written. It is true that the Allies would have lost all if they had been defeated in the west, and that the call of the Armies for more and more men and munitions for that theatre was insistent; it is equally true, however, that in France there could be nothing but batter and counter-batter, and the only remaining point where strategic principles could be brought to bear was at the Dardanelles. But what is more relevant to the subject of these pages is that when in future years the story of Helles and Anzac and Suvla is weighed, it will, I think, appear that had the necessary air service been built up from the beginning and sustained, the Army and Navy could have forced the Straits and taken Constantinople. I insistently urged the dependence of the naval and military forces upon air assistance and the necessity for carrying out a strong aerial offensive, especially by bombing, for which the local conditions governing the enemy operations on the Peninsula offered exceptional advantages.

From the autumn of 1915 onwards Egypt became the centre of training and expansion for operations in the Middle East and, as the organization developed, a brigade was formed with Wings in Macedonia, Sinai and a training Wing, which by 1918 had become a training brigade, in Egypt. The work of the Wing sent to Sinai in 1916, and expanded in 1917 into a brigade, is well summarized in the following extract from a telegram received from Egypt on October 3rd, 1918:--

"Before operations commenced our mastery of the air was complete and this was maintained throughout, enabling the cavalry turning movement to be completely protected and concealed. Enemy retreating columns were so effectively machine gunned and bombed by offensive machines that in all three cases the surviving personnel abandoned their vehicles and consequently upset all plans of retirement. An enemy column thus abandoned was seven miles in length."

The Wings in Macedonia and Mesopotamia, though they could not beat the record of the Palestine Brigade, gained a marked supremacy over the enemy. Air operations in East Africa were originally carried out by the Royal Naval Air Service with seaplanes, which in 1915 were brought up to the strength of two squadrons and replaced by aeroplanes under the orders of the military forces, their duties being carried out under the difficult conditions of bush warfare. Valuable work was also done by the Royal Flying Corps squadrons which were sent out to operate in the south.

In addition to these major operations, air forces were used in the expeditions on the Indian frontier, against Darfur and in the vicinity of Aden. Five squadrons were sent to Italy after the Italian retreat from the Isonzo and took a prominent part in the final Austrian defeat; a Royal Air Force contingent was sent to Russia to operate from Archangel; and material assistance was given to France and the other Allies, but especially to the United States in the training and equipment of her air forces.

At the beginning of 1918 the Royal Flying Corps and the Royal Naval Air Force were amalgamated and the Royal Air Force came into existence, and during the year achieved a supremacy more complete than that at any time since the Somme.

The following description gives a vivid idea of air activity at the front in 1918:-

"All day long there were 'dog fights' waged at heights up to three or four miles above the shell-torn battlefields of France, whilst the low-flying aeroplanes were attacking suitable targets from the height of a few dozen feet. Passing backwards and forwards went the reconnaissance machines and the bombers, and along the whole front observers were sending out by wireless to the artillery the point of impact of their shells. Such was the picture of the air on any fine day at the time."

1918, however, saw not only the accumulative effect of the tactical co-operation of aircraft with our armies in the field, but also the formation of the Independent Air Force and the carrying out of the strategic air offensive against centres of war industry in the interior of Germany.

A vast organization was also required at Home to meet the rapid expansion of units in the Field and to supply reinforcements. Thus at the Armistice there were 199 training squadrons, the pupils under instruction including cadets numbered 30,000, and during the war some 22,000 graduated as efficient for active service. At the beginning of the war pilots were sent overseas with only 11 hours' flying experience. This was much too little and there is no doubt that increased training would have ensured fewer casualties. Fortunately, however, the length of training was increased in the latter part of the war and a remarkable advance in training was made possible by the use of an entirely new and extraordinarily efficient system of instruction evolved by Smith-Barry.

The war demonstrated the beginnings of what air power meant, though in November, 1918, it was still in its infancy. Before many years the ability to make war successfully, or even at all, will depend upon air power.

Let us now briefly survey the development of the several duties of aircraft, the evolution of machines and progress in tactics, strategy and the organization of our Air Forces during the war.

I had recognized the great difficulty of mobilizing with the clockwork precision of older units and, in the belief that war was coming, had ordered a provisional mobilization of the Corps some days before it was actually declared. Thanks to this step and to the work done at our Concentration Camp at Netheravon in June, 1914, the greater part of the Royal Flying Corps was enabled to concentrate without hitch at our aerodrome at Dover, and the machines flew via Calais to Amiens on August 13th.

CO-OPERATION WITH THE ARMY.

Reconnaissance.

In the event of France and England declaring war concurrently against Germany, the strategic plan agreed to by the British and French general staffs before the war had been that the British Expeditionary Force should be moved to the Le Cateau, Maubeuge, Mons, area and take up a line on the left flank of the French Army near Mons. But England had withheld her declaration until three days after the French, and on landing in France the first words I heard said by a Frenchman were: "Oui, l'arm anglaise arrive mais on a manqué le premier plan." It was not until after the arrival of G.H.Q. at Amiens on August 14th that, although late, it was decided that the advanced line should be taken up. The Royal Flying Corps moved by air and road to an existing aerodrome outside the antique defences of Maubeuge 12 miles from Mons on the 16th. On the 19th the first reconnaissance was carried out, and the entire country over which the German armies were advancing, as far as Brussels and Louvain, was kept under observation. One of the best reconnaissances ever made was that of August 21st, which discovered the 2nd German Corps moving from Brussels through Ninhove and Grammont.

From Maubeuge we had to retire on the 24th to Le Cateau, on the 25th to St. Quentin, on the 28th to Compiene, on the 30th to Senlis, on the 31st to Juilly, on September 2nd to Serris, on the 3rd to Touquin, on the 4th to Melun, where we were thankful at last to get orders again to advance on the 7th to Touquin, and on the 9th to Coulommiers, reaching Tardennois on the 12th for the Battle of the Aisne.

Of the many recollections of the early days one which will remain longest in my mind is the terrible sadness of the flocks of refugees, of the poor people we left behind. And the glare of villages burning by the hand of the Boche. It was indeed war.

Valuable reconnaissances were made during the whole Retreat from Mons to the Marne in spite of the tremendous difficulties involved by constant movement, transport, and the selection of new landing grounds, but, in the words of Sir John French, "It was the timely warning aircraft gave which chiefly enabled me to make speedy dispositions to avert danger and disaster. There can be no doubt indeed that even then the presence and co-operation of aircraft saved the very frequent use of cavalry patrols and detailed supports." The Royal Flying Corps was an important factor in helping the British Expeditionary Force to escape von Kluck's nearly successful efforts to secure another and a British Sedan.

The reconnaissance resulting in the most valuable information of all, and, I think, during the whole of the war, was that of September 3rd, during the critical operations on the Marne, which formed one of the decisive battles in the world's history, when von Kluck's turning movement to the south-east against the French left was accurately reported and Marshal Joffre was enabled to make his dispositions accordingly. "The precision, exactitude and regularity of the news brought in," he said in a message to the British Commander-in-Chief, "are evidence of the perfect training of pilots and observers." The reports of the German air service, on the other hand, would appear from von Kluck's movements to have been of no assistance to him.

The system adopted from the first was for the pilot or observer, or both, immediately on their return to bring their report to R.F.C. Headquarters, whence the Commander, or his staff officer, accompanied them to G.H.Q., where the map was filled in in accordance with the report. G.H.Q. could then ask questions and obtain any further information which the observer could give, while R.F.C. Headquarters could ascertain what further reports were most urgently required. The form of the reports, which were ready printed, had been most carefully thought out at R.F.C. Headquarters in peace and experimented with at the Concentration Camp.

The maps thus compiled at G.H.Q. from air reconnaissance reports between August 31st and September 3rd were of vital interest, though it was sometimes very difficult to get the information put on the map for prompt consideration. For instance, at Dammartin on the evening of September 1st, when it was thought that German cavalry were within a few miles, G.H.Q. made a very hurried departure, and I was unable to find anyone to whom to give very important reports.

It was at the Battle of the Marne that machines were for the first time allotted to Army Corps for tactical work, while long-distance reconnaissance was carried out by other machines operating from Headquarters. Later on, this system was established as a part of our permanent organization, squadrons being allotted to, and reporting direct to, Corps for tactical reconnaissance, artillery co-operation and contact patrols, and to Armies for longer-distance reconnaissance and fighting.

The last phase of the war of movement was the race for the Channel Ports and it devolved upon aircraft to observe the enemy's movements from his centre and left flank to meet the Allied movement to the coast, to observe the movements of the four newly-formed corps which came into action at Ypres and to maintain liaison with the Belgian and British forces at Antwerp and Ostend. Information was very difficult to obtain and on one occasion I flew from the Aisne to Antwerp, under Sir John French's instructions, in order as far as possible to clear up the general situation when our G.H.Q. was in doubt as to whether Antwerp was completely surrounded or not. It was an interesting piece of work. There was a light drizzle, and the forest of

Compiene had to be flown over at about 200 feet. The B.E. could not make the distance without refilling, and although only a short halt was made at Amiens for the purpose, it was too late to fly direct to Antwerp. Instead, a landing was made in a very sticky field under light plough, which was selected from the air about 4 miles north of Bruges, to which town I rode on a borrowed bicycle. At Bruges there was great consternation and uncertainty as to the position at Antwerp, but the Commander kindly placed a large open car and its very energetic driver at my disposal to try and get through. After many difficulties we managed to find our way into Antwerp by about midnight, and I was received by the Belgian Commander. He explained that though the Germans had broken through the South-Eastern sector and his troops were very hard pressed (and pointing repeatedly to a piece of an 18-inch German shell in the corner of the room, he said, "Mais qu'est-ce qu'on peut faire avec ces choses-l?"), he hoped to be able to hold out for a time. After giving him General French's message and obtaining as much information as possible, I managed to get clear of Antwerp, reaching Bruges again at 3.15 a.m. At 4 a.m. we set out and found a very wet machine in a wetter field and after considerable difficulty and flying through the top of the surrounding hedge, struggled into the upper air on the way back to Headquarters at Tardennois.

During the Battles of the Aisne and of Ypres strategical reconnaissance was carried out by the few machines available at Headquarters. Shephard, the best reconnaissance officer I have ever known, who was killed later, used to fly his B.E.2 without observer over the greater part of Belgium two or three times a week and always brought in a long, closely packed, and extraordinarily valuable report. Tactical reconnaissance to a depth of 15 to 20 miles was done by units attached to Corps.

After the Battle of the Aisne, which was the turning point in the evolution from the war of movement to trench warfare, pure reconnaissance, though still the basis of air work, tended to become a matter of routine, while many new and specialized forms of it--such as air photography and artillery spotting by wireless--were developed.

Photography.

Though experiments had been made in the problem of photography from the air before the war, principally by Fletcher, Hubbard and Laws, and its value to survey was recognized, it had not become of practical utility. We only took one official camera with us to France on August 13th, 1914, and it was not until September 15th that the first attempt at air photography was made, when five plates were exposed over positions behind the enemy's lines with very imperfect results. Its great value as an aid to observation in trench warfare was, however, very apparent, fresh brains were brought to the task, Moore-Brabazon, Campbell and Dr. Swan, and by the end of the year better success was obtained, though positions even then had to be filled in by the observer with red ink. Experiments at home during 1915 led to a great improvement in lenses, and at the beginning of 1916 air photography was universal. At the Battle of the Somme new enemy positions were photographed as soon as they were seen, and the camera did invaluable work in the reconnaissance of the Hindenburg Line during the German retreat of 1917, and the taking of over a thousand photographs was a daily occurrence. On September 4th, 1917, a record of 1,805 photographs was made.

The development of air photography, very remarkable in itself, is even more so when it is remembered that the improvement in enemy anti-aircraft guns drove our machines to carry out their work at altitudes increasing up to 20,000 and even 22,000 feet, at which heights the negatives had to be as distinct as those taken at 4,000 in the earlier days of the war.

At the beginning of the Dardanelles operations our apparatus consisted of one camera, a printing frame and a dark room lamp. The first photographs were taken by Butler in April, 1915, from a H. Farman machine at necessarily low altitudes. Butler was wounded in June and was succeeded by Thomson, who alone made 900 exposures and sent in 3,600 prints.

In addition to the assistance of air photography to reconnaissance, the war gave it great impetus as the handmaid of survey and mapping. It was, in fact, the only means of mapping or correcting the maps of country held by the enemy, which in certain cases, as at Gallipoli and in Palestine, were very inaccurate.

By the end of the war photographic processes and equipment had reached a high standard of excellence. There are still, however, certain difficulties in regard to the production of accurate maps, which have not been overcome, the most obvious being the necessity of an initial framework of fixed points and of contouring. The subject is considered so important that an "Air Survey Committee," consisting of representatives of the Air Ministry, the Geographical section of the War Office, the Ordnance Survey, the School of Military Engineering and the Artillery Survey School, has recently been formed. In addition, the School of Aeronautics of Cambridge University is studying the question. The Survey of India and the Survey of Egypt are also conducting experiments.

Wireless.

From the outset, part of the German scheme of tactics was to batter down resistance by means of superior weight of heavy armament, and with the beginning of warfare of fixed position the observation and direction of our artillery fire became as important as distant reconnaissance. Besides its immense value in increasing the effect of the batteries, it had the indirect advantage of more closely binding the ties of mutual understanding between the air and ground troops, a point which fortunately seems to have been misunderstood by the Germans. In September, 1914, the first attempts were made to signal enemy movements from the aeroplanes of a Headquarters Wireless Flight which had been formed for the purpose, and this practice was continued with success throughout the Battle of the Aisne.

In the earliest stages artillery co-operation was also carried out by dropping coloured lights, but from the Battle of Ypres onwards, though for some time very few wireless machines were available, this was effected by wireless or signal lamps. In his dispatch on the Battle of Loos, Sir John French wrote: "The work of observation for the guns from aeroplanes has now become an important factor in artillery fire, and the personnel of the two arms work in closest co-operation."

By the Battle of the Somme artillery co-operation had assumed very large dimensions. For instance, on September 15th, 1916, on the front of the 4th Army alone, seventy hostile batteries were located, twenty-nine being silenced. Counter-battery work was so effective before the offensive which opened on the Ypres front at the end of July, 1917, that the Germans withdrew their guns and the attack was delayed for three days in order that their new positions might be located.

Recognition marks on aeroplanes were at this time, and indeed throughout the war, a matter of great difficulty. It had been suggested before the war that they would not be necessary, but the reverse was found to be the case, as even with the distinctive marks which were adopted our machines were often fired at by British troops, and we should undoubtedly have lost very heavily if we had flown over our own lines with false marks, as was suggested, or none.

Bombing.

The bombing operations, which reached their climax in the raids on German industrial centres in 1918, arose from very primitive methods used at the beginning of the war. During the retreat from Mons a few hand grenades were carried experimentally in the pockets of pilots and observers, or, in the case of the larger varieties, tied to their bodies, and these were dropped over the side of the machine as opportunity occurred. At the Marne, for instance, small petrol bombs set fire to a transport park and scattered a mixed column of infantry and transport. I think I am right in saying that the first German bombs were dropped on us--unsuccessfully--at Compi 鐩 ne on August 29th, 1914. It was not, however, until the beginning of 1915 that special bombing raids were started by the Royal Flying Corps, one of the first places to be attacked being the Ghistelles aerodrome in West Flanders.

The most important bombing operations and raids into Germany in the early days of the war were carried out by the Naval Air Service, units of which landed at Ostend on August 27th and operated with the Royal Naval Division from Antwerp. They were subsequently withdrawn to Dunkirk to form the nucleus of an aircraft centre from which excellent work was done in attacking the bases established on or near the Belgian coast from which German submarines and airships conducted their operations.

Just before the Germans entered Antwerp, the first raid was made against a German town, one machine reaching Dusseldorf, when it descended from 6,000 to 400 feet and dropped three bombs on an airship shed.

From the end of 1914 onwards the activities of the Royal Naval Air Service in this theatre of operations continually increased, the chief objectives being the gun emplacements at Middelkerke and Blankenburghe, the submarine bases at Zeebrugge and Bruges, the minefield and dock of Ostend, the airship sheds near Brussels, and the dockyards at Antwerp. The first airship destroyed in the air was attacked over Ghent.

An interesting experiment was the attempt by the R.N.A.S. at the Dardanelles to sink the heavy wire anti-submarine net, which had been stretched on buoys across the Straits at Nagara by the Turks, by means of parachute bombs.

To return to the Royal Flying Corps. During 1915 railway junctions were the principal bombing objectives, and raids were carried out on an ever-increasing scale, formations of fourteen to twenty machines taking part. At the Battle of Neuve Chapelle for instance, the railway junctions at Menin, Courtrai and Douai were attacked. One officer of No. 5 Squadron, carrying one 100 lb. bomb, arrived over Menin at 3,500 feet, descended to 120 feet, and dropped his bomb on the railway line. The first V.C. of the Royal Flying Corps was obtained at the Second Battle of Ypres by Lieutenant W. B.

Rhodes-Moorhouse, who in bombing Courtrai came down to three or four hundred feet, under heavy fire, but piloted his machine 35 miles back to Merville at the height of a few hundred feet, and died a few days later from his wounds.

One of the most instructive features of the Battle of Loos in September, 1915, was the definite co-ordination of bombing attacks with army operations. Many types of machines, belonging both to Army and Corps Squadrons, carried bombs in order to destroy dumps, communications, cut off reinforcements, and the like, while at the Somme bombing was carried out by formations of Wings. In October, 1917, 113 tons, and for a period of six days in March, 1918, 95 tons, of explosives were dropped. This illustrates the enormous progress of bombing which was so largely resorted to in the later stages of the war. The hand grenades of 1914 had become bombs weighing three-quarters of a ton: the pilot's pocket a mechanically released rack: and aim, assisted by instruments, was becoming fairly accurate.

Night bombing, necessitated by the fact that by day a large machine heavily laden with bombs was an easy prey to the fighting scout, came into prominence in 1916, increasing in intensity up to the end of the war; and raids into Germany recommenced. Early in 1918 these raids included the bombing of Maintz, Stuttgart, Coblentz, Cologne, and Metz. Machines sometimes dropped their bombs from heights of about 12,000 feet and at other times descended to within 200 feet of their objectives.

Contact Patrol.

Contact patrol, the name given to the direct co-operation of aircraft with troops on the ground, was first extensively practised at the Battle of the Somme, though experiments in this direction had been made in 1915, messages being dropped at the Battle of Neuve Chapelle at pre-arranged points.

The main objects of contact patrols were to assist the telephone (which was frequently cut by shellfire), to keep the various headquarters informed of the progress of their troops during the attack, so also saving them from the possibility of coming under the fire of their own artillery, to report on enemy positions, to transmit messages from the troops engaged to the headquarters of their units, to attack ground formations, and to co-operate with tanks. A system of red flares on the floor of the trenches was used to mark the disposition of the troops, and aircraft communicated their information by means of signalling lamps, wireless and message-bags.

During the German retreat of 1917 contact patrols attacked enemy foundations from 100 feet and in some cases landed behind the enemy lines to obtain information. The skill of low-flying pilots in taking cover by flying behind woods, houses, etc., became increasingly important. The fact that 62,673 rounds of ammunition were fired from the air against enemy ground targets between November 20th and 26th, 1917, and 163,567 between March 13th and 18th, 1918, indicates the rapid development of this form of aircraft action, the effect of which was frequently more deadly than bombing.

Two of many protagonists of contact patrol were Pretyman and Bishop. On one occasion the latter, in attacking an aerodrome at about 50 feet, riddled the officers' and men's quarters with bullets, put two or three machines on the ground out of action, and three in succession as they got into the air. Another interesting example of contact patrol work occurred in 1917 when a pilot flew his machine at a low altitude over the enemy trenches, and he and his observer attracted the attention of the Germans by firing their machine guns and Verey lights. The Germans were so busy with the aeroplane that they had their backs turned to the front line and our infantry were able to cross no-man's land without any artillery preparation, take prisoners and bomb dug-outs.

An article in the Cologne Gazette showed what the Germans thought of low "strafing."

"The operations" (i.e. of June 7th, 1917), it says, "were prefaced by innumerable enemy airmen, who, at the beginning of the preparation for the attack, appeared like a swarm of locusts and swamped the front. They also work on cunningly calculated methods. Their habit is to work in three layers-- one quite high, one in the middle, and the third quite low. The English who fly lowest show an immense insolence; they came down to 200 metres and shot at our troops with their machine guns, which are specially adapted to this purpose."

Armour was first employed as a result of Shephard finding at Maubeuge a bullet lodged in the seat of his leather suit. Thin sheets of steel were at once cut out and placed in the wickerwork seats of aeroplanes. This primitive protection developed into the armoured machine mentioned later, which was about to make its appearance at the Armistice.

I may mention here the "special duty" flights, which consisted in establishing secret communication between our Intelligence Branch and agents in the territory occupied by the Germans. Agents, mostly French and Belgian, were carried by aeroplane over the enemy lines and landed there. This work was started in 1914.

Fighting.

At the beginning of the war it became obvious that it was not only the duty of aircraft to obtain information but also to prevent enemy aircraft crossing our lines. In addition to the reconnaissance machine, and in order to make its work possible, a machine designed purely for fighting was required. In August, 1914, the aeroplane's armament consisted simply of rifle, or carbine, and revolver, but our pilots nevertheless attacked hostile machines whenever the opportunity occurred. The first German machine to fly over us was at Maubeuge on August 22nd, 1914, and, though fighting on an extensive scale did not take place until 1916, as early as August 25th, 1914, there were three encounters in the air in which two enemy machines were driven down. One interesting report of an early fight is that between a B.E. and a German machine on December 20th, 1914.

"A German aeroplane with one passenger and pilot being encountered over Poperinghe, we followed to Morbecque and then to Armenties. The passenger of the B.E. fired 40 rounds from his rifle and the German passenger replied with some rounds from his revolver. The B.E. crossed the bows of the German machine to permit the pilot to use his revolver. The German switched off and dived below the B.E., and is believed to have landed somewhere north-west of Lille."

Another instance of the early air combats was when Holt, single-handed, and armed only with a rifle, lashed to a strut of his machine, attacked ten Germans near Dunkirk, causing them to drop their bombs in the field and make off to their own lines.

We managed to bring down a number of German machines, mainly by rifle fire (five had already been brought down by September 7th, 1914), but our great difficulty early in the war was to get the enemy into action, and, although during October and November, 1914, there was a certain amount of fighting, as a rule the German when attacked made for his own lines and the protection of his anti-aircraft guns. This, though offensive carried to the extent of wastefulness of life is equally bad, was a serious mistake in all ways from his point of view, entailing as it did a tendency for the confidence of the troops and the morale of the air service to be undermined from the outset. The error was rectified, but only temporarily, at the Somme.

As the specialized duties of aircraft increased, the Corps machines engaged in them needed protection and it was realized that the best method of protection was the development of the air offensive. This was rendered possible by the adaptation of the machine gun to the aeroplane. Early in 1915 the invention of the "synchronizing gear" enabled a machine gun to fire through the propeller, and by the end of 1915 fighting in the air became the general rule. The first squadron, No. 24, composed purely of fighting machines, took its place on the Western Front in February, 1916, and gradually Wings were attached to Armies solely for fighting and the protection of Corps machines. During the long months of the Battle of the Somme, for instance, when, though the Royal Flying Corps dominated the air, the Germans put up a strenuous opposition, bombing machines were protected by fighting patrols in formation on the far side of the points attacked. The rapidity with which fighting in the air developed is shown by the fact that at the end of 1916 twenty new fighting squadrons were asked for on the Western Front; the establishment was increased to twenty-four machines per squadron, and by the end of the war even night-fighting squadrons were operating with considerable success and, had the war continued, would have proved a very important factor in air warfare.

The development of aerobatics, air fighting, and formation tactics brought many airmen into prominence. For example Albert Ball, who ascribed his successes to keen application to aerial gunnery; J. B. McCudden, the first man to bring four hostile machines down in a day; and Trollope, who later on brought down six. Hawker met his death fighting von Richthofen, who describes the fight in his book The Red Air Fighter as follows:--

"Soon I discovered that I was not fighting a beginner. He had not the slightest intention to break off the fight.... The gallant fellow was full of pluck, and when we had got down to 3,000 feet he merrily waved to me as if to say, 'Well, how do you do?'... The circles which we made round one another were so narrow that their diameter was probably not more than 250 or 300 feet.... At that time his first bullets were flying round me, as up to then neither of us had been able to do any shooting."

At 300 feet Hawker was compelled to fly in a zig-zag course to avoid bullets from the ground and this enabled Richthofen to dive on his tail from a distance of 150 feet.

This indicates a heavy disadvantage under which our aircraft laboured in all their work on the Western Front. The prevailing westerly wind which, while it assisted the enemy in his homeward flight, made it very difficult for a British machine, perhaps damaged by anti-aircraft fire, to make its way--still under fire--to its base.

I cannot leave the subject of air fighting without giving one or two more examples. One which comes to mind is that of five British machines attacking twenty-five of the enemy. One of ours gliding down with its engine stopped and being attacked by two Germans was saved by another British one attacking and driving off the two enemy. The result of the combat was five German machines destroyed and four driven down out of control, whilst all of ours returned safely. Another example, that of Barker who, whilst destroying an enemy two-seater, was wounded from below by another German machine and fell some distance in a spin. Recovering, he found himself surrounded by fifteen Fokkers, two of which he attacked indecisively but shot down a third in flames. Whilst doing this he was again wounded, again fainted, again fell, again recovered control and again, being attacked by a large formation, shot down an enemy in flames. A bullet now shattered his left elbow and, fainting a third time, he fell several thousand feet, where he was again attacked, and thinking his machine had been set on fire he tried, as he thought in a final effort, to ram a Fokker, but instead drove it down on fire! Barker was by this time without the use of both legs and an arm. Diving to a few thousand feet of the ground he again found his retreat barred by eight of the enemy, but these he was able to shake off after short bursts of fire and he returned a few feet above the ground to our lines.

Though at the beginning our machines were rather better than either the French or German, it was the marked superiority of our pilots which gave us the greatest advantage. We should have been superior even had the machines been exchanged.

CO-OPERATION WITH THE NAVY.

We have seen that the functions of co-operation with the Navy--Coast defence and Fleet assistance--were very complicated, and that at the outbreak of war the splendid pilots and excellent equipment of the R.N.A.S. were not so highly organized and were wanting in cohesion, but that the

R.N.A.S. had advanced further than the Royal Flying Corps in specialized technical development. In the earlier part of the war, in addition to its main duties, the R.N.A.S. ventured in many directions, many of them of considerable value to the Army, as, for instance, at Antwerp.

Coast Defence, Patrol and Convoy Work.

Immediately war broke out a system of coastal patrols was established between the Humber and the Thames Estuary and over the Channel--the latter serving as an escort to the Expeditionary Force crossing to France. Patrols were at first, through limitations of equipment, mainly confined to the Home coast, but, as the war went on and machines improved, they were rapidly extended, especially in connection with the detection and destruction of submarines; reconnaissances were carried out over the enemy's shores, and in 1918 there were forty-three flights of seaplanes, thirty flights of aeroplanes, together with flying boats and airships, operating from, and communicating with, an ever-increasing number of shore stations. Not only was anti-submarine work carried out in the vicinity of the coast, but organized hunts were made for submarines, ships were convoyed on the high seas, shipping routes were protected, and action was taken to bar the passage of submarines through narrow channels. This was effected by an intensive system of combining and interlocking patrols, and by maintaining, in close co-operation with surface craft, a protective barrage across suitable stretches of water, such as the Straits of Dover.

Airships from the beginning, when patrols operated from Kingsnorth during the crossing of the Expeditionary Force to France, proved particularly useful for escort, in addition to patrol work, and twenty-seven small airships, known as the S.S. type, were completed in 1915. In 1916 the Coastal type with a longer range was designed and constructed and new airship bases were established.

Fleet Assistance, Reconnaissance, Spotting for Ships' Guns.

The successful use of Drachen kite-balloons borne in ships at the Dardanelles led to their extensive development. Up to about May, 1915, when the vessels to which they were attached could stand in close to shore and overlook the enemy's positions from a distance of three or four thousand yards, a large amount of spotting of great value was carried out by these balloons for ships at Gallipoli, but when the Turks brought long-range guns into position, kite-balloon vessels were obliged to lie out beyond 11,000 yards and their services were rendered comparatively slight for this purpose. From 1916, however, they were towed by merchant auxiliaries and light cruisers to spot submarines, observers communicating with the patrol ship by means of telephone. One of the most wonderful sights I have ever seen was from the observer's basket of the kite-balloon let up from S.S. "Manica" in June, 1915. We were spotting for the guns of H.M.S. "Lord Nelson" bombarding Chanak. The sky and sea were a marvellous blue and visibility excellent, the peninsula, where steady firing was going on all the time, lay below us, the Straits, with their ships and boats, the Asiatic shore gradually disappearing in a golden haze, the Gulf of Xeros, the Marmora, and behind one the islands of the 灾 ean affording a perfect background. No one who was at the Dardanelles, however vivid the horrors and the heat and dust and flies, will forget the beauty of the scene, especially at sunset, and it was seen at its best from the basket of a kite-balloon.

The ever-increasing assistance rendered by aircraft to surface vessels in crippling Germany's submarine campaign is shown by the fact that in 1915 ten submarines were attacked from the air and in 1918 126 were sighted and 93 attacked. Nor was the principle forgotten in countering the submarine menace that offence is the best defence, and among the many duties of R.N.A.S. aircraft, based on Dunkirk from the early days of the war, were anti-submarine patrols along the Belgian coast and the bombing of hostile submarine bases, such as Bruges.

As in the case of the Army Corps observation machines, fighting scouts became necessary for the protection of patrols and to counter the enemy's efforts at raids and sea reconnaissance, and the considerable amount of experiment in air fighting which the R.N.A.S. had made before the war bore useful fruit.

For the immediate protection of the Grand Fleet seaplane and aeroplane bases were established at Scapa Flow and Thurso at the beginning of the war, but, owing to damage from a gale in November, 1914, aircraft operations with the Fleet were carried out from the seaplane carrier "Campania." The problem of using carriers with the Fleet had not been seriously tackled before the war, and though experiments were strenuously carried out, and there were fourteen carrier ships in commission in 1918, and a seaplane carrier operated with the Battle Cruiser Squadron at Jutland, the use of aircraft in this way did not become very efficient. One of the chief difficulties was limitation in size, and consequently in radius of action, of aircraft employed from carriers or the decks of battleships. The total number of aeroplanes and seaplanes allotted to the Grand Fleet in 1918 was 350.

Seaplane carriers occasionally co-operated with fighting ships. For instance in October, 1915, a fast carrier at the Dardanelles accompanied ships detailed for the bombardment of Dedeagatch, and her seaplanes not only co-operated in spotting but also made a valuable reconnaissance of the Bulgarian coast and railway. But as a rule fighting and reconnaissance aircraft had mainly to work from shore bases. To assist in this direction, units were sent overseas to be nearer their sphere of action, as, for instance, the R.N.A.S. squadrons stationed at Dunkirk which, besides general reconnaissance, helped the Navy to keep open the Straits of Dover, carried out bombing raids against German bases and dockyards, such as Ostend, Zeebrugge, and Bruges, and co-operated with monitors in the bombardment of the Belgian coast. The development of a long-range seaplane or flying boat was also taken in hand, though an efficient type was not produced until the last year of the war.

As with the Army, an important part of naval aircraft duties was spotting for gunfire; and likewise single-seater fighters were required for the protection of our own aircraft, for preventing enemy aircraft reconnaissance, for attacking the enemy's fleet and protecting our own. The use of offensive patrols steadily increased during the war.

Bombing.

I have already referred to bombing and mentioned the attack on Dusseldorf as an instance of the work done. Bombing raids had always been looked on with favour by the R.N.A.S. and were used throughout the war as a means of countering hostile aircraft operations from bases in Belgium. One of the first successful raids was that against the Friedrichshaven Zeppelin works by three Avro machines, which flew 250 miles over enemy country on November 21st, 1914. Another noteworthy example was the attempted raid against Cuxhaven on Christmas Day, 1914, carried out by seaplanes, which were still in an experimental stage, and three carriers escorted by naval units. Powerful machines for bombing purposes were ordered and bombs of greatly increased size and gear for dropping them were designed.

Torpedo Attack.

The impetus given to bombing helped forward another use of naval aircraft: torpedo attack. This is likely to develop in the future into one of the most important uses of aircraft in naval operations, but during the war it was never given an objective by the German fleet. In May, 1915, two Sunbeam Short machines were embarked in the "Ben-my-Chree" for operations at Gallipoli, and it was in this theatre that for the first time in history ships were sunk by torpedoes released from aircraft. I shall never forget the night when we steamed silently up the narrow Gulf of Xeros and lay waiting to release our seaplanes in the still darkness of the early morning. The machines were lowered noiselessly into the water, and, their engines started, flew across the narrow neck of Bulair under fire from the old Turkish line; then, reaching the northern end of the Dardanelles at dawn, they descended low (one machine actually landed on the water and discharged its torpedo), sank their targets, and returned. In addition to the possibility of submarine attack, the Gulf of Xeros is so narrow that our ship could have been hit by the cross fire of field guns. It was a very fine performance and, although during many years I have spent anxious hours hoping for the distant purr of a safe returning machine, I have never been happier than when after a long wait our seaplanes were again quickly raised on board. The only torpedo machine employed at the Battle of Jutland was a Sunbeam fitted with a 14-inch torpedo, and it was not until just before the Armistice that a squadron of torpedo aircraft was ready for operations with the Grand Fleet.

The Germans also tried to develop the use of torpedo-carrying seaplanes and, as with their submarines, had the advantage over us of a vast number of targets close to hand in our North Sea and Channel shipping, but fortunately the British fighting scouts were able to destroy several of their machines before they had done much damage.

HOME DEFENCE.

At the beginning of the war the R.N.A.S. assumed responsibility for the defence of Great Britain against attacks by hostile aircraft, and a scheme for the defence of London and other large towns was entrusted to an anti-aircraft section of the Admiralty Air Department. Its resources, however, consisting of a few unsuitable and widely scattered aeroplanes, some 1 pdr. pom-poms with searchlights manned by a special corps, were inadequate and it was fortunate that only three small daylight aeroplane raids, mainly for reconnaissance, were made during 1914--the first German machine to visit England dropping a bomb near Dover on December 21st.

Night Flying and Night Fighting.

In spite of continuous action by the R.N.A.S. against German airship bases in Belgium, there were in 1915 nineteen airship and eight aeroplane raids--one by night--over England, and, although the new and powerful Zeppelin L.Z.38, which attacked London on May 31st, was destroyed by an aeroplane counter-attack in its shed near Brussels, no real counter measures were evolved until 1916, when Home Defence was taken over by the War Office. During that year a Home Defence Squadron of B.E.2c's, rapidly expanded to a Wing, was formed; and the systematic training of night pilots, the standardization of night-flying equipment and armament, and the lighting of aerodromes, was taken in hand. A continuous aeroplane and searchlight barrage with night landing grounds was gradually formed between Dover and the Forth; the wireless signals employed to assist Zeppelins in finding their way were intercepted, thus enabling our rapidly improving fighting machines to pick up and attack raiding airships; and the constant attacks to which airship sheds were exposed in Belgium, caused their withdrawal to positions further inland and increased their distance from England. During 1916 there were twenty-two raids by airships, six of which were destroyed, the first being brought down in September at Cuffley by Leefe Robinson. Thenceforward airship raids declined, the destruction of the majority of the largest and latest which raided England on October 19th, 1917, sealing their fate.

On the other hand, aeroplane daylight and night raids on London, the first of which occurred in November, 1916, increased in number and strength with the object, in addition to the destruction of material and civilian morale, of forcing upon us the unsound retention at home of a considerable air defence force. The largest of these attacks was made by seventeen aeroplanes at midday on June 13th, 1917, but, the Zeppelin danger nullified, counter measures to meet the new menace were gradually evolved. New squadrons were raised and the number of home defence squadrons was raised to fourteen service and eight night training squadrons; a Northern Home Defence Wing was formed at York; and the Home Defence Group became the 6th Brigade. The first night aeroplane raid occurred in September, and the systematic training of night-fighting pilots on scout machines was hurried on. Separate zones for aeroplanes, guns and searchlights--the latter provided with sound locators--forming an outer barrage, were instituted, and aprons, supported by kite-balloons, formed a protective barrage up to 8,000 feet. A system of wireless and ground telephonic communication was improvised for plotting the course of attacking aircraft and thus enabling squadron commanders to concentrate machines at the point of attack. By 1918 the night-fighting aeroplane, assisted by these means, had countered the night-bombing aeroplane. At first, this had been the result of the retention of a large number of fighting aircraft and a complete organization at home.

Meanwhile, night fighting, especially the protection of night bombers by fighting machines, had become of paramount importance on the Western Front. The chief feature of activity in September, 1918, was the successful co-operation between searchlights in the forward areas and No. 151 night-fighting squadron. This was the first night-fighting squadron, trained by the 6th Brigade, to be sent to France. It was proposed to send four more such squadrons and thus form a first line of offensive defence which would react on hostile raids over England. Thus once again the old doctrine was gradually observed that offence is the only true defence, and that purely defensive measures, however efficient, by keeping men and material from the vital point, are necessarily expensive out of all proportion to their effectiveness. Both the Germans and ourselves made the initial mistake of organizing large local defence systems partly to placate public opinion. During the German offensive of 1918 a further development of night fighting took place in the bombing and low strafing of enemy troops and unlighted transport with the aid of flares.

THE MACHINE AND ENGINE.

Turning now to the machine and engine, the Military Trials held in 1912, when the Royal Flying Corps was started, represented the first organized effort to assist the evolution of service aeroplanes in this country and a brief comparison will be useful to show the performance of the average machines and engines of that date, at the beginning, and at the end of the war, and of civil machines of to-day.

At the Military Competitions of 1912, of the eight types--Avro, B.E., Bristol, Cody, Bleriot, Deperdussin, Hanriot, and M. Farman--the first four were British, though only the Avro had a British engine, and the last four French, fitted with French engines. The average horse-power was about 83, the average maximum speed 67, and the minimum 50 miles per hour; the climb to 1,000 feet was effected in 4-1/2 minutes with an average load of 640 lb., which included pilot, fuel for four hours and useful load. The loading per square foot was, for biplanes, about 4-1/2, and, for monoplanes, 6 lb.

On the outbreak of war, and until the end of 1914, of the ten types in use--Avro, B.E., Bristol, Sopwith, Vickers, M. Farman, H. Farman, Caudron, Morane, and Voisin--five were British and five were French and all were fitted with French engines. The average horse-power was still about 83, but the average maximum speed had risen to 74, and the minimum had fallen to 41 miles per hour. The load averaged 609 lb.

A remarkable advance in machine and engine construction is shown by referring to the tables for 1918. At the Armistice of the twelve types--Avro, Bristol Fighter, Sopwith Snipe, S.E. 5a, de Havilland 4 and 9a, Vickers Vimy, Handley Page O/400 and V/1,500, Fairey Seaplane 3c, F. 2 A. and F. 5--all were British and, except the de Havilland 9a, which had an American engine, were fitted with engines of British manufacture. The F. 2 A., and F. 5, were twin-engined, while one, the Handley Page V/1,500, was equipped with four engines. The average horse-power was per engine, 344, and per machine, 516; the average maximum speed 111, and the minimum 53-1/2 miles per hour, the climb to 6,500 feet was carried out in 13 minutes and to 10,000 feet in 24 minutes with an average load, including fuel for 5-1/2 hours, of 2,742 lb. The average ceiling was 15,500 feet; the loading per square foot about 8 lb.

The years following the Armistice have witnessed the conversion of military machines and the development of new designs for commercial purposes. In 1921 there were thirteen types fitted with British engines: Avro, Bristol, de Havilland 4, 16 and 18, Vickers Vimy, Handley Page O/400 and W. 8, B.A.T., Westland, Fairey, Supermarine and Vickers Amphibians. No British machine had a foreign engine. The Vickers Vimy, Handley Page O/400 and W. 8, which had a passenger-carrying capacity of 15, were twin-engined. The average horse-power was per engine, 387, and per machine, 474; the average maximum speed 114, and the minimum 49, miles per hour. With an average load of 2,467 lb., including fuel for 4-1/2 hours, 19 minutes was required for a climb to 10,000 feet. The average loading per square foot was about 13 lb., and the average ceiling 15,793 feet.

Before the war, in addition to the Royal Aircraft Factory, there were only eight firms engaged, on a very small scale, in the manufacture of aircraft in England, and an aero engine industry hardly existed. Until 1916, the greater proportion of our machines, and almost all our engines, were French, and we were very dependent upon France for the replacement of our heavy losses in material. By the end of the war the bulk of our material was of British design and construction, though there was still a certain number of British built engines of French design. One American engine--the Liberty--was also employed. The fact that in October, 1918, the Royal Air Force had 22,171 machines and 37,702 engines on charge, and that during the ten months January to October the output of machines had been 26,685 and of engines 29,561, gives some idea of the enormous growth in production.

In the first few months of the war it was not possible to progress far with new inventions or improvements. Fortunately, our Aircraft Factory had evolved in the B.E. a machine of considerable stability which in this respect compared favourably with German machines, and was well adapted to its work of reconnaissance.

Technical progress during the war often unfortunately involved the loss of valuable lives, as for instance those of Professor Hopkinson and Busk, to both of whom heavy debts of gratitude are owed, but gradually obstacle after obstacle, problem after problem, was successfully tackled by our designers and constructors. With a view to enlarging the field of observation, staggered planes were introduced in the B.E.2c. This machine also proved that it was possible to calculate the degree of stability and thus paved the way for the design of aeroplanes with indifference to stability and increased man[oe]uvrability for fighting purposes, or with great inherent stability for bombing. During 1915 the B.E.2c was used for all purposes, but the extra loading involved by the increasing use of aeroplanes for bombing and fighting caused a decrease in the rate of speed and climb, and our aeroplanes were temporarily inferior in fighting power to the Fokker.

The necessity of preventing the enemy obtaining information soon led to the development of air fighting. At the beginning of the war the sole armament of aeroplanes was the rifle or revolver. The machine gun soon followed, but its use in tractor machines was impracticable on account of the danger of hitting the airscrew. The first "fighters" were therefore two-seater pushers, such as the "Short-horn" Maurice Farmans which, though not designed for fighting, and too slow to chase enemy aircraft, were the first to be fitted with Lewis guns, and F.E.'s, the first machine designed specifically for fighting, with the machine-gun operator in front of the pilot. These "pusher" fighters had an excellent field of view and fire forwards, but suffered from lack of speed and a large "blind" area to the rear. On the other hand, the single-seater tractors were potentially the superior fighters, and in order to protect the blades of the airscrew the French were the first to use deflector blades on them in tractor machines.

Our early single-seater tractors were fitted with a Lewis gun fixed so as to fire over or at the side of the airscrew and actuated by a bowden wire, the most efficient, though not the most numerous, fighting machines at the end of 1915 being the Bristol Scouts.

By the Summer of 1916, however, we had adapted the "synchronizing gear" to our machine guns, enabling them to be fired through the propeller; while aircraft engines developed much greater power and full allowance was made for all equipment carried. From that time the development of our single-seater fighters was steadily progressive. One of the first of these was the Sopwith "Pup," which had a speed of 106-1/2 miles an hour at 6,500 feet, climbed 10,000 feet in just over 14 minutes, and could attain a ceiling of 17,500 feet. In 1917 appeared the Sopwith "Camel," a typical example of this type, which was simple, stable, easily controllable and possessed two guns. It had a speed of 121 miles an hour at 10,000 feet, to which height it could climb in under 10-1/2 minutes, and a ceiling of 23,000 feet. The Martinsyde F.4, embodying further improvements, was not ready in time for active service.

While the single-seater tractor was developing for purely offensive action, the two-seater fighter, of which the field of view, man[oe]uvrability and general performance were being improved, retained its utility as a reconnaissance machine. In 1916 the "pusher" type was superseded by the Sopwith "1-1/2 Strutter" armed with a synchronized Vickers gun, which for its 130 horse-power was never surpassed. The pilot was close to the engine and had a good view of the ground, while the gunner was placed behind him with a rotary Lewis gun turret. Early in 1917 these qualities were further developed in the Bristol Fighter.

With the advent of these improved types the B.E.2c was relegated to the work of artillery co-operation, until superseded by the B.E.2e. Towards the end of 1916 appeared the R.E.8 with a Vickers synchronized gun and a Lewis gun, which after many vicissitudes became the standard machine for artillery work.

Systematic bombing was practised by nearly all types of machines, but real accuracy was never obtained. Thus, the B.E.2c was first used in formations, but with a full load of bombs it could not carry an observer, and its moderate speed left it an easy prey to hostile fighters. Early in 1916 appeared the Martinsyde single-seater bomber with an endurance of 4-1/2 hours, and in 1917 the D.H.4 which was much used for day-bombing. The F.E.2b pusher, discarded as a fighting machine, became the principal night-bomber.

It was comparatively late in the war before special bombing machines were evolved. They were then divided into day-bombers and night-bombers, the D.H.9 and 9a machines being typical of the former and the Handley Page of 1917--a large twin-engine aeroplane, the first really effective night-bomber, of considerable carrying power but low performance--of the latter. By November 8th, 1918, two super-Handley Pages were ready to start to Berlin. They possessed a maximum range of 1,100 miles, a crew of seven, four 350 horse-power Rolls-Royce engines, arranged in pairs, a tractor and a pusher in tandem on either side of the machine, and, as they would be compelled to fly both by night and day, a gun defence system. The D.H.10a and the Vickers Vimy, for day and night bombing respectively, were also being produced at the date of the Armistice.

In the early days of the war an aeroplane had little to fear above 4,000 feet. With the improvement of the anti-aircraft gun there was, by the end of the war, no immunity at 20,000 feet. Very low flying for attack was, however, being rapidly developed, and would have proved of great effect in 1919. The aeroplane used for this purpose was the single-seater fighter, and the Sopwith "Salamander," with two guns, a speed of 125 miles an hour, and 650 lb. of armoured plates, was about to make its appearance at the Armistice.

I have previously mentioned how dependent the improvement of design and performance of aircraft has been upon the less simple and tardier development of the engine. The invention of the light motor made aviation possible, and development has synchronized with the evolution of lighter, more powerful and more reliable engines. One of the most difficult problems still confronting us is the production of a cheap, high-powered and reliable engine, but the existence at the end of the war of machines weighing 15 tons indicates the progress achieved, while British engines of 600 horse-power are now in use, and one of 1,000 horse-power will shortly be available.

TACTICS AND THE STRATEGIC AIR OFFENSIVE.

During the war there were three concurrent movements in process: the ratios of the various forms of air tactics were constantly changing, and the components of our air forces varied in accordance with the development of reconnaissance, artillery co-operation, bombing and fighting. Secondly, their total strength was increasing rapidly; and, thirdly, it was increasing relatively faster than the Army or Navy.

It was an evident and logical development and in accord with the shortage of national man power and the consequent tendency to a reduction in the strength of the Army, that, the necessary uses of aircraft with the Army and

Navy being ensured, any available margin of air power should be employed on an independent basis for definite strategic purposes. The difficulty was to arrive at an agreement as to the minimum tactical and grand tactical requirements of the Army and Navy. The British Army was not alone in asserting that there was no minimum and that it wanted every available airman, and agreed with the French that anything which it could temporarily spare should be lent to the French Army. It was argued that the Armies could as easily and better arrange for strategic bombing. Fortunately in 1918, when I was Chief of the Air Staff, we managed to secure a margin and formed the Independent Air Force in June of that year. It was, of course, understood that, in the event of either the British or French Armies being hard put to it, the Independent Air Force could temporarily come to their direct assistance and act in close co-operation with them.

In 1915 in accordance with the old doctrine that offence is the best defence, the surest method of protecting specialized machines on the battle front was found to be in the attack of enemy aircraft by fighting machines. In 1918 it was decided that raids on the centres of German war industry would not only cripple the enemy's output of material essential to victory, but also relieve the pressure on the Western Front, the vital point of the war. The Germans had had the same intention in the many raids which started over Dover on December 21st, 1914.

Long-range bombing had, however, been carried out spasmodically before 1918. In addition to its taste for bombing in general, the Royal Naval Air Service were keenly bent from the outset on long-range bombing in particular. The question of forming an Allied squadron to bomb German munition factories was first raised in 1915 at one of the monthly meetings between the French and British Aviation departments; and in February, 1916, a small squadron of Sopwith "1-1/2 Strutters" was formed at Detling for the purpose of bombing Essen and Dusseldorf from England, but the Army in France, being short of machines, asked that they should be sent to the front, and therefore the scheme did not mature; neither, for similar reasons, did one for the co-operation in 1916 of British and French bombing squadrons, operating from Luxeuil.

It was not until October, 1917, that the first striking force, consisting of three squadrons, was formed under the Army with Ochey as its base. It was mainly used in raids against the ironworks in the Alsace-Lorraine Basin and the chemical industry in the neighbourhood of Mannheim. As I have said, a definite offensive policy by means of an independent strategic force was later decided upon, and the "Independent" Air Force was brought into existence. It originally comprised two day-bomber and two night-bomber squadrons. During the summer additional squadrons were allotted to it, including D.H.9's and Handley Pages. Day-bombing squadrons had to fight their way to objectives in close formation, and the problems connected with navigation, calculation of petrol supply, action of wind and ceiling, were all accentuated. Casualties were heavy, with the result that a squadron of Fighters, composed of Sopwith "Camels," was incorporated for the purpose of protection. Thus we see the beginnings of an air fleet analogous to the naval fleet with its capital ships and protective craft.

The main objectives were the centre of the chemical industry at Mannheim and Frankfort; the iron and steel works at Briey and Longwy and the Saar Basin; the machine shops in the Westphalian district and the magneto works at Stuttgart; the submarine bases at Wilhelmshaven, Bremerhaven, Cuxhaven, and Hamburg, and the accumulator factories at Hagen and Berlin.

It will be seen from a map that three of the main industrial centres were situated near the west frontier of Germany; and, therefore, one portion of the striking force was based at Ochey, which lies within a few miles of the Saar Basin, within 180 miles of Essen, and within 150 miles of Frankfurt. Another portion was based on Norfolk, where a group of super-Handley Page machines were established for the specific purpose of attacking Berlin, a distance of 540 miles, and the naval bases within 400 miles. It was obvious that though aircraft from England would have to cover greater distances, they would not expose themselves to the strong hostile defences in rear of the battle front.

Three instances of the Independent Air Force's action may be cited. On the night of August 21st/22nd, two Handley Page machines dropped over one ton of bombs on Cologne Station, the raid occupying seven hours. On the night of August 25th/26th two Handley Pages attacked the Badische Aniline und Soda Fabrik of Mannheim; bombs were dropped from a height of 200 feet, direct hits being obtained in every case; and the machines then remained over the town, which they swept with machine-gun fire. On August 12th the first attack was made on Frankfurt by twelve D.H.4 day-bombers, every machine reaching the objective and returning safely in spite of being attacked, over Mannheim and throughout the return journey, by some forty hostile fighters.

During the five months of its existence the Independent Air Force dropped 550 tons of bombs, 160 by day and 390 by night. Of these 200 tons were dropped on aerodromes, largely by the short-distance F.E.2b's, as a result of which, hostile attacks on Allied aerodromes became practically negligible. Theoretically, machines of the Independent Air Force should not have been utilized for attacking purely military objectives in the Army zone, such as aerodromes, and their co-operation with the Army for this purpose shows that their true role was either not appreciated or not favoured by the French and other Commands.

There is ample testimony to the spirit of demoralization which pervaded the civil population of the towns attacked.

"My eyes won't keep open whilst I am writing," reads one captured letter. "In the night twice into the cellar and then again this morning. One feels as if one were no longer a human being. One air raid after another. In my opinion this is no longer war but murder. Finally, in time, one becomes horribly cold, and one is daily, nay, hourly, prepared for the worst." "Yesterday afternoon," says another, "it rained so much and was so cloudy that no one thought it was possible for them to come. It is horrible; one has no rest day or night."

Although, for reasons into which it is not necessary to enter here, only a comparatively small percentage of the efforts of the Independent Force were directed against the industrial targets for which the force had been created, yet by the end of the war the strategic conception of air power was bearing fruit, and the Air Ministry had in hand measures for bombing which would have gone far to shatter German munitionment. The defence measures forced upon the Germans within their own country were reacting on their offensive action at the front, which was at the same time denuded of fighting aircraft at various points to meet the menace of our strategic force at Ochey.

ORGANIZATION.

As in peace on a small, so in war on a large scale, the history of the organization of aircraft, while we were fighting for our national existence and competing with similar enemy expansion, is one of continuous development, of decentralization of command and co-ordination of duties. Headquarters, the Squadron and the Aircraft Park, as originally conceived in peace, though subject to variations in size, remained the basis of our organization. For instance, the original eighteen machines of our squadron were increased to twenty-four for single-seater fighters and reduced to six in the case of the super-Handley Page bombers. The four squadrons originally operated directly under Headquarters, were soon allocated to Corps for tactical reconnaissance and artillery co-operation, while a unit remained at Headquarters for strategical and long-distance reconnaissance and a few special duties. The next step was in November, 1914, when two Wings, composed originally of two, and later, of five squadrons each, were formed, R.F.C. Headquarters retaining one squadron and the wireless flight for G.H.Q. requirements. The Wing Headquarters co-ordinated the work of the squadrons which were allocated to Army Corps.

A further development, in 1916, was the formation for each of the three Armies of a Brigade, consisting of two Wings and an Aircraft Park. One--the Corps Wing--carried out artillery co-operation and close reconnaissance (including photography) with Army Corps, the other--the Army Wing--carried out more distant reconnaissance and fighting patrols under Army

Headquarters. Our air superiority at the Battle of the Somme in 1916 led us to expect German counter-measures in 1917, and our programme for the following winter contemplated a proportion of two fighting squadrons to each Corps Squadron. By 1917 there were five British Armies in France and Belgium and our air forces were increased to provide a Brigade for each of the two new Armies. The Headquarters of the flying force in the field (except in the case of the Independent Air Force, which was responsible to the Supreme War Council and the Air Ministry in London) remained attached to G.H.Q. throughout the war.

The main difficulty in the higher organization was the lack of co-operation between the Royal Flying Corps and the Royal Naval Air Service and their competition for the supply of men and machines--the demands of both being urgent and insatiable. As a first step to overcome this, an Air Board was formed in May, 1916, to discuss general air policy, especially the combined operation of the Naval and Military Air Services, to make recommendations on the types of machines required by each, and to co-ordinate the supply of material. The Air Board was an improvement, but not a remedy, and, therefore, in 1917 it was decided to form an Air Ministry responsible for war aviation in all its branches and to amalgamate the Naval and Military Air Services as the Royal Air Force. This was carried into effect early in 1918, with Lord Rothermere as Secretary of State for Air with a seat in the Cabinet, and the air became the third service of the Crown, with an independent Government department permeated with a knowledge of air navigation, machinery, and weather, and closely allied to the industrial world for the initiation, guidance, and active supervision of research and experimental work.

I will mention later some of the many arguments for and against the retention of an independent Air Ministry and autonomous Air Force in peace. The amalgamation was certainly advantageous in war. It effected the correlation of a number of hitherto independent services according to a uniform policy and prevented overlapping by centralizing administration. Under single control it was possible to carry out, on a carefully co-ordinated plan, recruiting and training, to supply men and material, to organize air power according to the strategic situation in each of the various theatres of war, and to form the correct ratio between the air forces in the field and the reserves in training at home. The difficulty was that the amalgamation had to be carried out during the most intensive period of air effort, but by the end of the war most of these objects had been attained without jeopardizing the close co-operation with the Army and Navy. Co-operation with the Naval and General Staffs and with naval and military formations was, in fact, improved, independent action was beginning to bear fruit, and we possessed an Air Force without rival.

CHAPTER III

PEACE

THE FUTURE OF AERIAL DEFENCE.

In the evolution of aviation during the war the conclusion has been reached that the most remarkable lines of development at the Armistice were in the direction of ground and night fighting, torpedo attack and long-range bombing, exemplifying respectively the three spheres of air operations-- military co-operation, naval co-operation, and the strategic use of aircraft. It must be remembered that this progress in tactics and strategy, in the machine, and the airman's skill, was made in the short period of four years, and that every war has started with a great advance in scientific knowledge, accumulated during peace, over that obtaining at the close of the previous war. We may therefore assume, provided the danger is averted of a retrograde movement from recent scientific methods to pre-war conditions-- sabres, bayonets, and guns--that by the outbreak of another war on a large scale, which we hope may never occur, the knowledge of Service aeronautics will have increased immeasurably since 1918, and may be, not a contributory, but a decisive factor in securing victory.

The period since the Armistice has been employed in the reduction and consolidation of the Royal Air Force. In England the cadre system has been adopted, while abroad the greatest concentration of effort is aimed at, with Egypt, at present the most important strategic point in the Imperial air system, as the centre of activity. Iraq is being handed over to the control of the Royal Air Force, whose share in the policing of overseas possessions is likely usefully to grow provided any tendency to the concurrent building up of a large ground organization is withstood. The advantages of aircraft for "garrison" duties lie, under suitable geographical conditions, in their swift action and wide range, their economy, and, during disturbances their capacity for constant pressure against the enemy without fear of retaliation. One of the main problems is at present that of personnel. Service flying is restricted to comparatively young men, and therefore the majority of officers can only be commissioned for short periods. For this reason the experiment is being made of taking officers direct from civil life on short engagements, and at the same time endeavouring to ensure, by technical and general education, that the Royal Air Force shall not become a blind-alley occupation.

Though it is difficult to foretell on what lines aircraft will develop for any one purpose, as in the past, the problem of military co-operation will perhaps be less complex than that of co-operation with the Navy. It will probably consist of improvements along the lines already indicated, such as increased range, speed, climb, man[oe]uvrability, offensive armament, armour, the assistance of tank and anti-tank action, and the utilization of gas. Fighting will undoubtedly take place at very high altitudes to keep the enemy's fighting machines away from the zone of operations--necessitating the development of the single-seater so as to increase climb and man[oe]uvrability, and obtain, if possible, a speed of 200 miles an hour at 30,000 feet. Cavalry, unless retained, as I think they should be, in the form of mounted machine-gunners, will, I think, disappear in European warfare, but infantry will remain, and it will be the object of aircraft to assist their advance by reconnaissance, ground attack, artillery and tank co-operation, and the destruction of the enemy's supplies and communications. In this connection ground tactics and air tactics must develop pari passu and commanders of Corps and Armies must work out during peace training the fullest schemes for the most intimate co-operation between air and land forces.

The future of naval co-operation is a difficult problem, more especially as there was no major naval engagement after Jutland in which aircraft could be used, and consequently we have little to go on in estimating their practical value in direct co-operation with the fleet. It is impossible at present to judge between the conflicting opinions as to the future of the capital ship, but it is certain that aviation will materially modify naval tactics and construction. Coast defence, reconnaissance, anti-submarine work, escort, and the bombing of enemy bases, will doubtless continue and develop with ever-increasing machinery and equipment; but torpedo attack by aircraft may reach a point where the very existence of opposing fleets may be endangered. It is already questionable whether a battleship could survive an attack launched by even a small force of this mobile arm.

As was the case during the war, the action of aircraft at sea is restricted by range, the difficulty being to find the mean between the opposing conditions of radius of flight and limitation in the size of aircraft imposed by the deck-space of "carriers," but there is reason to suppose that on the one hand engines will be so improved as to afford a sufficient radius of action to comparatively small aircraft, while, on the other, devices will be found to economize deck-space.

Fleets operating near the enemy's coast will be vulnerable from land aircraft bases, and thus close blockade will be rendered increasingly difficult. The possibility of gas attack on enemy bases from the air in co-operation with submarines and of effecting a blockade by this means must be envisaged.

Since the Armistice the operational work of the Royal Air Force on behalf of the Navy has been conducted under the auspices of the Admiralty. Improvements have been made in large flying boats and amphibians, especially with a view to facilitating their landing on "carriers" and the decks of battleships. There has also been considerable progress in the construction and use of torpedo aircraft.

The war lasted long enough to prove the effect of the strategic offensive by air. In spite of the dictates of humanity, it cannot be eliminated. It is true that modern war is inimical to the progress of mankind and brings only less suffering to the victors than to the vanquished. To ensure peace should therefore be our ideal. But a great war once joined is to-day a war of peoples. Not only armies in the field, but men, women, and even children at home, are concentrated on the single purpose of defeating the enemy, and armies, navies, and air forces are dependent upon the application to work, the output of war supplies, and, above all, the morale of the civil population. Just as gas was used notwithstanding the Hague Convention, so air war, in spite of any and every international agreement to the contrary, will be carried into the enemy's country, his industries will be destroyed, his nerve centres shattered, his food supply disorganized, and the will power of the nation as a whole shaken. Formidable as is the prospect of this type of air warfare, it will become still more terrible with the advent of new scientific methods of life-destruction, such as chemical and bacterial attack on great industrial and political centres. Various proposals, such as the control of the air effort, service and civil, of all countries by the League of Nations, and even the complete elimination of aviation, have been put forward as a means of avoiding the horrors of aerial warfare and its appurtenances, but they are untenable, and any power wishing and able to sweep them aside will undoubtedly do so.

A future war, as I see it, will begin something after this manner, provided either side possesses large air forces. Huge day and night bombers will assemble at the declaration of war to penetrate into the enemy's country for the attack of his centres of population, his mobilization zones, his arsenals, harbours, strategic railways, shipping and rolling stock. Corps and Army squadrons will concentrate in formation to accompany the armies to the front; reconnaissance and fighting patrols will scatter in all directions from coastal air bases to discover the enemy's concentrations and cover our own; the fleet, whatever its nature, will emerge with its complement of reconnaissance and protective machines and torpedo aircraft for direct action against the enemy's fleet. A few fighting defence units will remain behind.

But it must not be imagined that these functions will be carried out unopposed. Local battles in the air will occur between fighting machines for the protection of specialized machines, while the main air forces in large formations will concentrate independently to produce, if possible, a shattering blow on the enemy and obtain from the outset a supremacy in the air comparable to our supremacy on the sea in the last war.

In mobilization the time factor is all-important. Our national history has been one of extraordinary good fortune in this respect, but the margin allowable for luck is becoming very narrow and, whereas in 1914 it was some twenty days between the declaration of war and the exchange of the first shots, in the next war the air battle may be joined within as many hours, and an air attack launched almost simultaneously with the declaration of war. In modern war the mobilization period tends to shorten, and every effort will be made towards its further reduction, since mobilizing armies are particularly vulnerable from air attack.

CIVIL AVIATION AS A FACTOR IN NATIONAL SECURITY.

The picture I have drawn may appear highly coloured for the reason that no country is likely for some time to possess sufficiently large air forces to obtain a decisive victory, or at any rate an uncontested superiority, at the outbreak of war. Though in air, as in every other form of warfare, attack is more effective than defence, we cannot afford to keep our air forces up to war strength in peace any more than our Army or Navy.

The problem, from a military point of view, is therefore to ensure an adequate reserve and to maintain our capacity for expansion to meet emergencies. The number of units maintained at war establishment should be the absolute minimum for safety and of the type immediately required on mobilization, i.e. long-range bombing and naval reconnaissance squadrons.

The remainder should be in cadre form. We can, of course, maintain a fixed number of machines and pilots in reserve for every one on the active list, but, although some such system is necessary, on a large scale it is open to many and serious objections. First of all, even on a cadre basis, it means keeping inactive at considerable cost a number of machines which may never be used and which, however carefully stored, quickly deteriorate. Knowledge of aeronautics is still slender and improvements are made so continuously that machines may become obsolete within a few months. Moreover, the growth of service aviation in peace must tend to become artificial and conventional rather than natural, and this will react on design and construction, which will be cramped, both technically and financially, within the limits imposed by service requirements.

It is obvious therefore that the capacity of the construction industry to expand cannot be fostered by service aviation alone; furthermore, in the event of another war of attrition, expansion will be more essential than any amount of machine reserve power immediately available, and in the event of a war of short duration that power will win which has the greatest preponderance of machines, service or civil, fit to take the air. The asphyxiation of a large enemy city, if within range, can be done by night-flying commercial machines, and it would require a defending force of great numerical superiority for its successful defence.

Whether, therefore, from this point of view, or others, which I will mention later, another solution must be found, and this lies in the development of civil aviation. An analogy in the Navy and the Mercantile Marine has long been apparent. "Sea power," says Mahan, "is based upon a flourishing industry." Substitute "air" for "sea" and the analogy is still true. The Navy owed its origin to our mercantile enterprise and to-day it depends upon the Mercantile Marine for its reserve power of men and material. In the same way must air power be built up on commercial air supremacy. If we accept Mahan, or the dictum of any other great naval or military historian or strategist, a service air force by itself is not air power, and after a brief if brilliant flash must wither if reserves are not immediately at hand. A large commercial air fleet will provide, not only a reserve of men and machines, but it will keep in existence an aircraft industry, with its designing and constructional staffs, capable of quick and wide expansion in emergency; and such an industry will not be employed on the design of contrivances for use in a possible war, but on meeting the practical requirements of everyday air transport and navigation.

Thus a natural, practical and healthy, as opposed to a stereotyped and artificial, growth will be ensured. Our naval supremacy is largely attributable to the interest which the people as a whole have traditionally taken in naval policy; in other words, to the fact that we are a seafaring nation. Similarly air supremacy can only be secured if the air-sense of the man in the street is fostered, and aviation is not confined to military operations, but becomes a part of everyday life. At the present time commercial aviation is far too small to play the part of reservoir to the Royal Air Force--an object which must constitute one of the principal claims for support of the nucleus already in existence.

CIVIL AVIATION AS AN INSTRUMENT OF IMPERIAL PROGRESS.

Civil aviation, however, has not only an indirect military, but, with its superiority in speed over other means of transport, a direct commercial utility. The nation which first substitutes aircraft for other means of transport will be more than half-way towards the supremacy of the air. Moreover, as the Roman Empire was built upon its roads and as the foundations of the British Empire have hitherto rested upon its shipping, as steam, the cable and wireless have each in turn been harnessed to the work of speeding up communications, so to-day, with the opening of a new era of Imperial co-operation and consultation, this new means of transport by air, with a speed hitherto undreamed of, must be utilized for communication and commerce between the various portions of the Empire.

A comparison of the French and British attitudes towards civil aviation clearly demonstrates the two policies I have mentioned. Both France and England grant subsidies--France the very much larger sum--but the great difference lies in the objects aimed at. French policy is fostering civil aviation as a part of its military policy and, a portion of the subsidy being given to machines fulfilling service requirements, there is a strong tendency for French civil aviation to be military air power camouflaged. British policy, on the other hand, should aim at fostering civil aviation primarily as a commercial concern and believes that air commerce is the basis of air power as a whole. We are prepared to face the tendency of military and civil machines to diverge if that divergence is essential to the commercial machine.

An alternative to the British policy of maintaining a small air force and fostering commercial aviation as a reserve is the Canadian plan of a small air force training school and a civil Government flying service with such objects as forest patrol, survey and coastguard duties, the work being carried out on repayment for Government departments, provincial governments and private corporations. The former method, allowing of independent commercial expansion, is better suited to British mentality and requirements, but its success will depend on a genuine endeavour to make commercial aviation the real and vital basis of our air power. Experience in commercial operation cannot be gained by the exploitation of air routes or the carriage of mails or passengers under Service auspices. It is only by running transport services, as far as possible under private management, that operational data can be obtained, economies effected, and the design of strictly commercial machines improved.

To sum up. Military air supremacy can best be assured by the intensive development of industrial air organization for commercial purposes. The conception of civil aviation as the mainstay of air power as a whole is right. Service aviation is bound by technical and financial limits; its scope confined to the requirements of war. Civil aviation, on the other hand, opens out a prospect of productive expansion. The steady growth of the Continental services is already beginning to demonstrate the importance of air transport.

FINANCIAL AND ECONOMIC PROBLEMS.

The commercial exploitation of air transport is passing through a period of experiment, and suffering in the general war reaction from the incapacity of the public to think of aviation except as a fighting service. The machines hitherto used on the lines to and on the Continent are principally converted war machines, and to transform military into commercial craft and to use them as such is of small assistance to civil aviation, which requires reliable, economic machines as one of the basic conditions of its financial success. The cost of running an air transport service is considerable. Depreciation is one heavy item of expenditure. New machines must be evolved suitable to the requirements of mail, passenger and freight transport, but, in the present state of financial stringency, capital is not forthcoming for experiment unless there is every promise of a safe return. Then there are the expenses involved in general ground organization, maintenance, fuel, insurance, etc. The question is how can we carry on until the really economic type of commercial machine is evolved. It will never be evolved unless there is continuous flying and a continuous demand for new and improved machines for commercial work. To meet this in France, the Government came forward with a liberal grant of subsidies which have now been increased and placed on a more favourable basis, permitting of a very considerable reduction in the fares for transport by air. The British Government has also granted a subsidy for British firms operating on the cross-Channel routes, which it is hoped will place them before long on a sound, self-supporting, commercial basis. Part of this subsidy is allocated to assist transport companies in obtaining the latest type of commercial machines on a hire purchase system. With a few services properly supported by the State we shall pull through the experimental period of civil aviation.

The services to the Continent, although the distance is on the short side for the merits of air transport to be properly demonstrated, effect a considerable saving in time, and it is certain that the amount of mail, especially parcels, carried on these routes will continue to increase and lead to the eventual adoption of normal rates for air postage. An extension of the use of aircraft as the regular means of carrying mails will be of great assistance in the development of air transport. Aircraft revolutionize the speed of intercommunication by letter, and banks and financial houses will gradually realize that large savings can be made by utilizing air mails for the transaction of business. A difficulty lies in the fact that the area of the British Isles is not very favourable for an extensive air mail service, which can only operate by day, since by the existing means of transport mails are carried during the out-of-business hours and can generally reach their destination in a night, while the distances to Paris and Brussels are too short to afford outstanding advantage.

Lastly, we require public support and a spirit of confidence in the air. This can only be secured by increased reliability, reduction of charges and keeping the public informed of the progress made. It is the nature of man to distrust new departures. He disliked the introduction of mechanical devices into the Lancashire weaving mills. He scoffed at the steamship and railway. To-day he is inclined to treat as premature the serious exploitation of the air. In spite of the great decrease of accidents, in spite of the increased comfort of air travel, in spite of increased regularity, the average person is slow to realize that the communication of the busy man of the future will be by air. The majority of the business world is too conservative to make general use of the opportunities offered by aircraft for the quick transmission of its correspondence, while, though speed must be paid for, the high fares hitherto charged have deterred the general public from substituting the aeroplane for the train or boat. The running costs represented by these fares are being materially reduced as a more economic machine is evolved, and the reduction of fares which helps to place competition with foreign subsidized services and with the older forms of transport on more equal terms must for a time depend upon the assistance of Government grants.

WEATHER CONDITIONS AND NIGHT FLYING.

The safety of the machine and the reliability of an air service largely depend on accurate weather forecasts. In order to co-ordinate the meteorological work of the country as a whole, and for the special assistance of aviation, the Meteorological Services of Great Britain have been amalgamated under the

Department of Civil Aviation, and, working in close co-operation with the Communications Branch of the Department, have made improvements in the rapid collection and distribution of meteorological information for all purposes. In addition to the forecasts issued four times daily, collective reports are issued hourly by wireless from the London terminal aerodrome at Croydon and copies are distributed to transport companies and others concerned.

A feature of meteorology which is often overlooked is its economic value. By making use of a knowledge of the wind at different heights, aircraft can complete journeys more quickly than would otherwise be possible, and thereby save their own fuel and their passengers' time. This will be specially useful in the tropics where the regularity of the surface winds has its counterpart in the upper air, but even in Europe time-tables can be drawn up with due attention to the favourable and unfavourable effect of prevailing winds. The planning of airship routes in particular, must be considered in close connection with this aspect of weather conditions.

To-day, however, the aeroplane may be considered as an "all-weather" craft, save for mist and fog--the enemies of all transport and particularly to that of the air--to which unfortunately England is particularly liable during the winter. Experiments have been carried out on the dispersal of fog, the illumination of aerodromes by fog-piercing lights, and instruments to record the exact position of the aeroplane and its height above the ground, but success has not yet been achieved.

Similar to the problems of flying and landing in mist and fog is that of night flying. Until night flying is practicable, only half the value of the aeroplane's speed is obtainable, since other transport services run continuously day and night. Further, as machines become rapidly obsolete owing to technical progress, it is essential that they should be in use for the greatest number of hours during their life. Much has been done in the lighting and marking of aerodromes and in the equipment of aeroplanes with wireless telephone and direction-finding apparatus.

It may here be mentioned that there are two methods of obtaining the position of aircraft by means of wireless telegraphy, known as direction-finding and position-finding. Direction-finding is effected by means of two coils set at right angles in the aircraft, by means of which the bearing of a transmitting ground station with reference to the aircraft's compass can be taken. When two or more bearings on different ground stations, whose position is known, have been obtained, a "cut" or "fix" of the aircraft is obtained. The position-finding system consists of two or more ground stations fitted with apparatus capable of taking bearings with respect to true north and connected by direct telephone line. The aircraft calls up by wireless one of these stations, requests her position and then makes a series of signals for about half a minute. The stations take the aircraft's bearings, plot its position, and transmit the information to the aircraft. Wireless direction and position-finding, as well as wireless telephony, have on several occasions proved their value to navigation, but in spite of instances of successful night flying, developments have not been such as to render night services practicable.

Marine experience has been a valuable guide, but aerial illumination has entailed many new problems of its own--the distribution of light through very wide angles, the installation of light and powerful lamps in aircraft, the elimination of shadows and the prevention of dazzle, the provision of apparatus to indicate the strength and direction of the wind, and the like.

Very shortly the first organized and equipped night-flying route will be available; that between London and Lympne on the Continental air highway. The Boulogne-Paris section will probably be ready a little later. There will be four lighthouses on the English section, of which two will be automatic, requiring no attention for twelve months at a time. These, and many other, facilities will much assist the progressive establishment of services during the hours of darkness, and will provide valuable data for the establishment of other night-flying routes. There is no real difficulty given a reasonably clear atmosphere.

ORGANIZATION.

I have mentioned the broad lines on which the organization of the air services was built up before and during the war. We have seen that the initial foundations and framework remained and bore the great systematic structural development which was gradually required. In August, 1914, there were some 240 officers, 1800 men and 200 machines; in November, 1918, 30,000 officers, 170,000 men, and 22,000 machines, all of them better and of a higher performance than those of 1914. Our casualties during the war were about 18,000; air formations had been active in some fifteen theatres of operations; 8,000 enemy machines and 300 observation balloons had been destroyed; some three-quarters of a million photographs taken over hostile country, and 12,000,000 rounds had been fired from the air at ground targets. At Home two organizations had expanded independently from the same seed until, impeding one another's growth, their trunks had joined and a single and improved tree was the result.

This is the only country where a unified air service has been adopted. In war the arrangement was successful. Against its continuance in peace the Army and Navy urge that, with the best of wills, there is a great difference between having an integral branch of a service to work with other services and having to deal with an independent organization, and argue increased cost, duplication, competition and disjointed action. There is no doubt that the liaison of the General, Naval and Air Staffs must be closened, and if co-operation with the senior services was really becoming less satisfactory, a return to the old system should be considered amongst other alternatives, but I do not think that it should be so. It must also be remembered that, although air co-operation is vital to naval and military operations, it is fortunately unlikely that there will be another war for a long time and, meanwhile, the growing essential, independent strategic action would be irretrievably impaired by the reabsorption of the Air into the Army and Navy.

On the other hand, even apart from supply, such a reversion would also cause much duplication, e.g. training. The solution and the correct and logical outcome of the unification of the Air service is the close grouping of the three arms in a Ministry of Defence, and this, even in face of the obvious practical difficulties, should be adopted and co-ordination thus increased step by step. Apart from Supply, some of the services in which this could be effected are the medical, education, chaplains, mobilization stores, transport, works and buildings, accounting, communications, ordnance and national factories. A modified scheme might also be studied in which, under a Ministry of Defence, the Army and Navy each had tactical air units of seconded personnel for artillery co-operation, spotting and reconnaissance, and the Air Ministry dealt with supply, research, initial training and reserves, civil aviation and an independent air force.

One of many good examples of the necessity of co-ordination is afforded by the position of the aircraft supply services at the beginning of the war and their development. We have already seen that there were some eight private firms manufacturing aircraft in a small way and there was practically speaking no engine industry at all. For the Royal Flying Corps, the War Office had relied largely on the Royal Aircraft Factory, and, although the methods of control adopted had many advantages, there was in them a tendency to retard private enterprise and development. The Admiralty, on the other hand, had assisted by dealing almost entirely with firms for Royal Naval Air Service supply. The conditions in France fortunately were very much better than those in this country, and for the first year or two French factories helped us out with both machines and engines. By the end of the war we had the largest and most efficient aircraft industry in the world. There were no less than seventy-six great factories turning out vast numbers of complete aeroplanes, in addition to thirty-three manufacturing complete engines and over 3,000 turning out spares and equipment. Such expansion is not possible within a few weeks, it took a long time to arrive at this position, and it causes one very seriously to think what would have happened had France not been our ally, and points the moral which has been mentioned of the necessity for a thriving aircraft and engine industry in peace. During the war Germany also had a very large number of firms engaged on this work.

THE MACHINE AND ENGINE.

The general differences between service and civil requirements in aircraft fall under the headings of ceiling, load and speed. For service purposes very much higher ceiling and greater climb and speed are required and the design is much affected by the condensed nature of the load. For peace purposes, besides the primary advantage of speed which the air has over other forms of transport, regularity must be ensured and the correct ratio between speed, duration and load-carrying power determined. Great ceiling, man[oe]uvrability and climb are not required.

However great the speed and load, there is no value in air transport, whether for passengers or mails or goods, unless it is safe and also compares favourably from an economic point of view with the older methods. Without these the public cannot be expected to utilize air transport, nor is there any inducement to surrender mails and freight for carriage by air. Every endeavour compatible with economy is made, as far as the equipment of aerodromes and the organization of the routes are concerned, to render air navigation as safe as possible, yet, though both safety and economy of running have been improved, much remains to be done. Safety depends largely on engine reliability, fire prevention and the capacity of the machine to land in small spaces.

Though neither roads nor rails have to be laid and aircraft possess the great advantages of mobility and point to point transit, the initiation and maintenance of an air service is a very complex and costly matter. The utilization of converted war machines is no longer sufficient and those specially designed for commercial work are beginning to make their appearance. Such are the Handley Page W.8, the Vickers, the D.H.18 and 34, and the Bristol 10-seater.

The first two are twin-engine and the last three single-engine machines. Opinions differ as to the relative advantages of the twin and single-engine type. The first and running costs of the single engine are lower, but the twin has greater power and carrying capacity, while most pilots prefer to have a surplus of power over and above that required for normal flight. For these reasons, and because of the psychological effect on insurance companies and on passengers, the twin engine will probably remain in use for large commercial machines, until long-lived and economic engines of more than 500 horse-power are available. On the other hand, where extra power is not required, the twin-engine is not safer than the single-engine machine; no existing twin-engine commercial aeroplane can maintain its height and land safely with only one engine running. Experiments have been made, especially in Germany, on the multi-engined machine with all the engines in the fuselage, but its advantages have so far been counterbalanced by loss of efficiency due to transmission gearing and shaft drives to the propellers and the vibration and weight of the gearing.

High-powered engines are very expensive to run and every effort has therefore to be made by aerodynamic efficiency to carry more useful load with less horse-power. Improvement is being made in this direction; thus the D.H.18 carries eight passengers at 56 horse-power per passenger, the D.H.32 is designed for the same number at 45 horse-power each, and the D.H.34 for ten passengers at 45 horse-power each.

The two best German commercial machines, the Junkers and the Fokker, have a comparatively low horse-power and a low fuel load, but greater attention has been paid to the design of the machines, which are monoplanes with cantilever wings, offering less resistance to the air than our biplanes. One of the most difficult problems is to evolve a high-lift wing which does not impair the aircraft's speed in the air. For commercial machines we must aim at the largest possible commercial load, the smallest possible fuel load and, consequently, an engine which uses fuel economically and, conversely, a lighter fuel. The development of the engine is receiving constant attention, as are also various safety devices, among which may be mentioned those guarding against fire and those varying the lift of wings so as to lower the landing speed and thus decrease the dangers attendant upon forced landings.

In addition to the high initial cost of machines and engines, their maintenance also requires the greatest care. Detailed investigation must be made into all serious accidents. This is now compulsory under the new Air Navigation Act, and the fitness of pilots is ensured by periodical medical examination.

Apart from the weather, the safety of an aircraft depends upon its engine, and perhaps even more upon the installation and accessibility of engines and their adjuncts, such as the petrol, oil, water and ignition systems. During the earlier stages of the war the average life of an engine before complete overhaul was necessary was, of stationary engines, from 50 to 60 hours, and of rotary engines, about 15 hours. To-day these figures stand at 200 hours and upwards and from 50 to 60 hours respectively. For commercial purposes this must be further increased to 300-500 hours as a normal working period.

There are two schools of thought with regard to the efficiency, reliability and the economy of engines. One school advocates using a light power plant per horse-power, run normally at about half its maximum; the other favours a plant of greater weight, more solid construction and greater efficiency, running at nearly its full horse-power. The former is more expensive in primary cost and upkeep, but allows a higher performance and provides reserve horse-power for emergency; the latter is cheaper, but involves a certain risk owing to lack of surplus power. We have hitherto shown a tendency to adopt the former method, the Germans the latter. For commercial purposes a compromise will probably be found to be best.

Apart from the initial outlay on "air stock," the maintenance, overhead, fuel, insurance and depreciation charges are very heavy. These are much affected by such items as simplicity of design, strength against wear and tear, ease of assembly and interchangeability of parts, easily removable engines, increase in durability by the use of metal construction for parts of the machine and the propeller, the elimination of rubber joints, substitution of air for water cooling, facilities for loading and unloading in a commercial machine, simple and efficient navigational instruments and self-starter. Every improvement, however small, will assist to reduce running costs. Then revenue must be increased and the comfort of passengers, as, for instance, ventilation, warmth, luggage capacity and, more than all, a reduction of noise has to be carefully considered or they will not travel a second time by air. An effective engine silencer is at last well on the way. It is obvious what a great advantage this attainment will be both for service and civil purposes. Roughly speaking, a high-powered engine without a silencer is audible at a distance of some seven miles and at a height of 13,000 feet at night time, though these distances are reduced by about a third by day when normal ground noises exist. The bulk of noise is caused by the exhaust, the propeller and mechanical noises in the engine.

I cannot leave this subject without emphasizing the value of research, both abstract and concrete. But, though it is the keystone of progress, its results must largely depend on the amount of flying done. It is clear that for economic reasons new designs can only thoroughly be tried out by commercial use, and therefore again that real progress is dependent on commercial activity.

The advance of civil aviation is bound to be slower than was that of war aviation. But, as war experience improved old and evolved new types, so will peace requirements and experience shape the type and design of aircraft and engine best suited to its purposes. Although a good deal has under the circumstances already been achieved in peace, much remains to be done. Gradually, however, with a modicum of research, improvements in the factors already mentioned and the reduction of initial cost and maintenance expenses, air transport for mails, passengers and goods will take its place as a normal commercial public utility service, and the increased speed of communication will assist in the general development of trade.

AIR SERVICES: BRITISH, CONTINENTAL AND IMPERIAL.

International civil flying commenced officially on August 26th, 1919, and gradually expanded, both in the United Kingdom and on the Continent, especially during the summer of 1920. France, aided by considerable subsidies, conducted services from Paris to London, Brussels and Strasburg, from Toulouse to Montpelier and across Spain to Casablanca in Morocco; Belgium, from Brussels to London and Paris; Holland, from Amsterdam to London; Germany, in spite of the restrictions placed upon her, entered the field as a competitor and her aircraft flew regularly from Berlin to Copenhagen and Bremen, and from Bremen to Amsterdam. On the American Continent, the United States Post Office ran mail services from New York to Washington, Chicago, and San Francisco, with extensions from Chicago, St. Paul, Minneapolis, and St. Louis.

For reasons which I shall give, there were no internal services in the United Kingdom, but there were four companies operating air lines from London to Paris, one of which held the contract for the carriage of mails. There were also air mail services between London and Brussels and Amsterdam. The mileage flown and the number of passengers and the weight of goods carried were considerable, while the number of letters steadily increased, especially on the Amsterdam service; and an efficiency of 76 per cent., 94 per cent., and 84 per cent. was obtained on the London-Paris, London-Brussels, and London-Amsterdam services respectively.

It must be remembered that these results were obtained without any direct assistance on the part of the State, such as was given by the French Government to air-transport companies in the form of subsidies. British economic policy is traditionally opposed to subsidies, believing that enterprise can be healthily built up on private initiative. Therefore, until 1921 civil aviation had to content itself with the indirect assistance of the State, which consisted mainly in the adjustment of international flying; the laying-out and equipment of aerodromes on the air routes; the provision of wireless communication and meteorological information; research and the collection and issue of general information concerning aviation.

This indirect assistance, however, proved inadequate to maintain the progress achieved during 1920, and therefore the maintenance of air services by means of temporary direct financial assistance had to be arranged.

I have already pointed out the difficulty against which commercial aviation has to contend in regard to the geographical features and position of the United Kingdom. Its comparatively small size, the propinquity of industrial centres, our efficient day and night express railway services, especially those running north and south, lessen the value of aircraft's superior speed and militate against the operation of successful internal air services. Possible exceptions might include amphibian services between London and Dublin, accelerating the delivery of mails five or six hours; between Glasgow and Belfast, where the Clyde and the harbour of Belfast could be used as terminals; or between London and the Channel Islands. I may point out in parenthesis that the development of alighting stations on rivers passing through the centres of towns is important, as a great deal of time is at present wasted in reaching the aerodromes necessarily situated some miles outside large centres of population.

Our immediate opportunities of development near home are therefore afforded by the air services to Paris, Brussels, and Amsterdam; but even here the saving in time is not great, and our position is unfavourable compared to that of the United States, where the Post Office saves two days in the delivery of mails by air between New York and San Francisco; or compared to that of Germany, where Berlin is within a 350-mile radius of Copenhagen, Cologne, Munich, Warsaw, and Vienna, which is itself in an advantageous situation as the junction for a South European system extending to the Balkan States and the Near East.

The ultimate use of the air, however, is not exemplified by a few passengers flying daily between London and the Continent any more than by a few squadrons of fighting craft. In a decade or two overhead transit will become the main factor in the express delivery of passengers, mails, and goods. It is the one means left to the Empire of speeding up world-communication to an extent as yet unrealized. For the price of a battleship a route to Australia could be organized, the value of which would be beyond computation.

The British Empire as a whole offers vast fields for expansion. In Africa, Canada, and Australia are found the great distances suitable to the operation of aircraft, the wide undeveloped areas through which air transport may prove more economic than the construction of railways, and the trans-oceanic routes over which travel by steamship has reached, and in many cases passed, its economic maximum speed. Air transport, careless whether the route be over land or sea, unhampered by foreign frontiers, gives the Empire precisely those essential powers of direct, supple, and speedy intercommunication which ship and rail have already shown us to be vital.

Here again the geographical position of England presents a difficult problem. England is divided from the rest of the Empire by a wide expanse, either of ocean or foreign territory. Egypt, the starting-point for air routes to India, Australia, and South Africa, may be described as the centre of a circle of which England is on the circumference; and it may be some years before an aeroplane can complete the journey between England and Egypt with only Malta as a stopping-place.

The future of long-distance oceanic air routes may depend upon the airship. Lighter-than-air craft, mainly for reasons of cost and vulnerability, did not receive such an impetus from the war as did the aeroplane, but the modern airship has claims for use over distances exceeding 1,000 miles. It can fly by night with even greater ease than by day; fog is no deterrent; engine trouble does not bring it down; and it can take advantage of prevailing winds. It would reduce the sea journey from England to Karachi from 22 to 5 days; from England to Johannesburg from 21 to 7 days; and from England to Perth from 32 to 10-1/2 days. Its achievements have already been considerable. In November, 1917, the German L.57 flew from Constantinople to East Africa and back--a distance of 4,000 miles--in 96 hours; in June, 1919, the R.34 flew from East Fortune to Danzig and back in 57 hours; and in July it crossed the Atlantic, was moored out in America for four days, and returned, a total distance of 8,000 miles, in the flying time of 108 hours for the outward and

75 hours for the homeward journey.

Before and during the war Germany gained wide experience in the design, construction, and handling of airships. It is probable that as soon as the peace terms and financial position permit she will begin to establish this form of transport on a commercial basis. In accordance with the Peace Treaty, and the Ultimatum of the London Conference of 1921, the construction of aircraft of all kinds is at present forbidden, but Germany is fostering airship development by the means left at her disposal. Her scientists are probing the constructional problems connected with large airships, while efforts are being made, by financial and other assistance, to maintain her technical staffs and airship bases in existence. At the same time German commercial interests are negotiating with foreign countries with a view to the development of airships abroad, and plans are being discussed for an airship service between Spain and Argentina.

The United States, France, and Italy are all interesting themselves, either financially or constructionally, in the future of airship development.

In Great Britain we have made great strides, particularly in the construction of small types, and our practical air experience in lighter-than-air craft, during the war, is the greatest in the world. With a view to carrying out the experiments necessary further to demonstrate the capacity of airships for commercial long-distance flights, a few months ago the Department of Civil Aviation took over all airship material surplus to service requirements. The main object was to test the practicability and value of mooring airships to a mast. Up to the present, a principal factor militating against the economic operation of airships has been the large and expensive personnel required for handling them on the ground, especially in stormy weather. The mooring-mast experiments have had considerable success and airships have been moored in high winds and over long periods with the assistance of a very small personnel.

The Government has decided, however, though recognizing their potentialities for speeding up communications between the various Dominions and the Mother Country, that the operation of airships cannot be carried out by the State on account of the present financial position.

Recognizing the limitations of Home services and those to the Continent, it was for the purpose of directing attention to the Imperial aspect of civil aviation that the great demonstration flights were organized in which Alcock flew the Atlantic in a Vickers "Vimy," Scott crossed to the United States and back in the R.34, Ross-Smith flew from England to Australia, and van Ryneveld from London to the Cape.

These flights necessitated, too, considerable ground organization in laying out aerodromes, as the following report on one in Africa vividly illustrates: "If aerodromes are left unattended for one year," it says, "practically all the work would have to be undertaken afresh, particularly in Rhodesia. The growth of vegetation is enormous, especially during the rains, and grass will grow to a height of eleven feet in six months; and trees stumped two feet below the surface will throw out suckers and replant themselves within a month after the rains have started.... It is most important that rough drains should be traced.... I have just started planting Doub grass. This grass gives an ideal surface for landing, kills other grasses, and possesses deep interlacing roots which will bind the entire surface of the aerodromes, making it permanent and free from washaways and the formation of sluits."

The demonstration flights, however, showed what could, rather than what should, be done, and what we look for to-day is the inception of practical undertakings, however small, in the various portions of the Empire. The most important of these is the service contemplated between Egypt and India; another instance is afforded by the West Indies, which suffer from the lack of inter-island communications, both for mails and passengers, and this could be partially rectified by an air service employing seaplanes or amphibians for the Leeward and Windward Islands and the Bahamas, and between the Bahamas and the American Continent, where an American company is actually conducting a service. Another project, given up owing to recent disturbances, was one for a flying-boat service on the Nile. Services are also being considered from Malta to Italy, Geraldton to Derby in Western Australia, Sydney to Adelaide and Brisbane, and Melbourne to Hobart in Tasmania. Canadian activity takes the form of work carried out by Government-owned civil machines in connection with forest patrol, photographic survey, exploration, anti-smuggling patrols, etc. It would be a great advantage if railway and steamship companies seriously considered the value of supplementing their services by air.

With regard to Government undertakings on the Imperial air routes, Malta is being equipped with an aerodrome, and a line of wireless stations has been established between Egypt and India, but the organization of this route has been delayed owing to the recent disturbances in the Middle East, and the financial outlay involved in ground organization. As I have said, the air route on which we should first concentrate, over and above the Continental services, is that between Egypt and India. Both strategically and commercially it is the most important in the Imperial system; it is a step towards Australia; it offers possibilities of the greatest volume of traffic; it should be much simpler to control than many international routes, which inevitably have many complications; weather conditions are not unfavourable; and the time taken for the journey by sea would be reduced by about one-half. If the shortcomings in point of distance of the continental routes in reaping the full advantages of travel by air, and the importance of the best possible communications for the Empire, are recognized, it is essential that a practical form of assistance should be given in the near future to the conduct of weekly or even bi-weekly services each way between Cairo and Karachi. Although it will not be a commercial proposition for some time, the Egypt-Karachi route, shortening as it will the delivery of mails between England and India by two-thirds, and England and Australia by one-third, offers greater results than the various other schemes at present contemplated. There are, however, certain considerations which will have to be weighed before the immense amount of work necessary to its initiation as a commercial air route is begun. The French, for instance, hope to push a trunk air route to India via Constantinople, and this line has the advantage of avoiding a long sea and desert crossing. On the other hand, it will be a very difficult matter to negotiate the mountains of Anatolia.

If enterprises of this kind are successfully started, if each of our self-governing Dominions and Colonies encourages civil aviation within its own territory, and develops the air-sense of its people, each portion of the Empire, by a process of natural expansion, and by the gradual extension of local air lines to merge with those from other portions of the Empire, will assist in eventually forming a continuous chain of inter-Imperial air communication. Such a process of internal development, supported by close co-operation between the States of the Imperial Commonwealth, is the best method of obtaining rapidity of air intercommunication and a system of Imperial air bases necessary to the strategic security of the Empire.

CONCLUSION

Within the necessarily narrow limits of this survey there has been traced the history of aviation from the earliest days; the tremendous impetus given to it by the war has been described, during the course of which not only did air co-operation become essential to the Navy and Army, but the importance of the Air Force as a separate arm, with its own strategic action, steadily grew; the increasing preponderance which aerial warfare will have in the future, and the horrors which it may bring, have been touched upon; and the possibilities of civil aviation in peace and war have been outlined.

The conclusion has been reached that we cannot dispense with aviation, even if we would. We must consider it as a whole and lay down the broad principles on which it should be developed. The air (I write as one who during the last months of the war held the post of Chief of the Air Staff) materially helped, if it did not actually win, the fight. It has greatly complicated and increased the problems of defence. In future its influence on these problems will be still greater. The air has no boundaries. Great Britain and the Empire are no longer protected by the seas. A correct assessment of their needs will entail a growing ratio of air force to Army and Navy, and air power will in itself depend on the development of civil aviation.

But though air action may be expected with justice to grow in proportion to that of the Army and Navy, and will certainly absorb certain functions of both, it would be unwise, at this early stage of development, for air forces to attempt too much at a time--such as, for instance, to garrison geographically unsuitable countries.

A certain amount of reliance could also be placed on civil machines temporarily borrowed for purely policing measures in uncivilized countries, or for the assistance of Government during civil disturbances; and for such purposes it should not be difficult to devise a scheme, especially when the State exercises a measure of control through the grant of subsidies, for the obligatory enrolment of civil commercial pilots in the reserve, and for periodical refresher courses for pilots, who are not actually in the service of companies, at civil aerodromes. Such systems are in force in France and Canada. In the event of war the independent striking air force could thus count upon a large proportion of civil reserve pilots and machines.

Air, allied to chemistry and the submarine, will be a difficult combination to withstand. The more its potential terrors are grasped, the less likely is war to be loosed upon the world, and it cannot be realized too clearly how much more easily than any other instrument of warfare aircraft and gas can be cheaply and secretly prepared by a would-be belligerent. Meanwhile, if civil aviation can be built up as a productive organization to a position relative to that held by our mercantile marine, we must understand that it will ensure air supremacy better than a large unproductive outlay on armaments. And I am convinced that, with public support, this can, and will, be done. Others will do it if we do not. But air power, although drawing its vitality from the expansion of air commerce and the growth of the civil aircraft industry, must at the same time rely upon the nucleus of a highly trained and technical air force. Service aviation must be the spearhead, civil aviation the shaft, of our air effort.

The present isolation of England in terms of air from the rest of the Empire, and the geographical conditions already described, certainly render the national expansion of aviation, both external and internal, a difficult problem. It is clear that for this reason it must rather develop on an Imperial basis. The Dominions have already started valuable civil air work and have appointed Air Boards. Whatever the political settlement of Egypt may be, it is important that our air interests at this "hub" of Imperial aviation should be safeguarded. Air communication between the various portions of the Empire may prove of inestimable value in a future world war, and Dominion air forces may be able quickly to concentrate against enemy territory which is out of the range of aircraft operating from home. We have seen the value of aircraft operating from land bases for naval patrol, anti-submarine action, and direct attack on enemy shipping. With the increasing radius of action of seaplanes and other naval aircraft, the Army and Navy may be relieved of certain of their duties in coast defence and in protecting Imperial trade routes. For these reasons, aircraft bases are required throughout the Empire, and it is the commercial development of aviation which is the best means of ensuring their establishment. It will be for the Imperial authorities, while attending to local conditions and requirements, to co-ordinate as far as possible the air effort of the Empire, so that in peace communications may be developed and in the event of war its full power may rapidly be utilized on a co-operative basis.

Civil aviation is not, however, merely a method of amplifying service air power. It has a vast potential value of its own. Communications shape human destinies. The evolution of our civilization bears strongly the marks of the systems which at various stages have made the intercourse of men and ideas possible. Its history is one of endeavour to extend the limits imposed upon human living and mobility in each of the great phases through which it has passed.

There was the phase of the coracle and the roller-wheeled vehicle, stretching back into the roadless mists of unrecorded time; of roads which gradually linked the important areas of the Roman Empire; of inland and coastal waterways; of ocean traffic, and its huge advance with the discovery of steam-power, which brought England to the fore.

With each phase the world shrinks smaller and the mists of the unknown recede. The development of human mobility is the greatest marvel of the present age. We can hardly realize that it was only the other day, as these things go--in 1819, just a hundred years before the same feat was accomplished by air-that the first sailing ship fitted with auxiliary steam (and not until 1828 that a real steamship) crossed the Atlantic.

Strain and competition are increasing. Trains vie with ships; motor transport with trains. Telephones, wireless, cables, and flying are speeding up communications to a degree undreamed of a few years ago. If the air is to be a prime factor in the world-phase to come, how will the British Empire be affected? Stretching from Great Britain to Australia and the Pacific Ocean, the Empire depends more than any other political and commercial organization on the most modern and speedy communications, and as each of its portions assumes greater responsibility there is greater need for co-operation, the distribution of information, and the personal contact of statesmen and business men. "The old order changeth, yielding place to new"; and in communications the new order is air transport.

Equally important is the international aspect. To-day we are deeply concerned with the maintenance of peace, and this can be achieved, not so much by the action of Governments, or the efforts of the League of Nations, as by the personal association of individuals of one nation with those of another, and an increasing recognition of common interests. I conceive that civil aviation, by reducing the time factor of intercommunication, will tend to bring peoples into closer touch with each other and will make for mutual understanding. The Peace Treaty provided for an Air Convention for the international control of civil aviation. The Convention has been signed by all the Allied nations which took part in the war, and I hope other countries will shortly be included. As soon as the Convention has been ratified, the International Commission of Air Navigation will be established, and for the first time the world will see the international control of a great transport service. I believe this will prove an important practical step towards international co-operation and goodwill.

We have no excuse for ignorance of the effects of Imperial and international co-operation. The war gave us an example of what the British Empire can do, provided its combined knowledge and effort is brought to bear for one great purpose; and in no respect was this better exemplified than in the utilization and scientific development of aviation. The world-position of the Empire as a whole is still the best. Commerce and communications are its bonds, and, if we are so determined, it is in our power to shape the destinies of the future.

A definite advance has been made since the Armistice and, if all goes well, a very much greater one will be made during the next two or three years, and in ten years mercantile air services will be operating on a self-supporting basis. The science and concentration employed in the war must be made to serve the requirements of peace. Readiness for, and success in, war are vital when war is unavoidable, but in peace it is civil and commercial activity which is vital.

As in its infancy it seemed incredible to those responsible for the direction of the older services that the air would be their most valuable partner; as, during the war, they grudged its logical development to strike widely where they could not reach, and tried to tether it closely to them, so now in peace the air is struggling to attain the apotheosis of communication.

In the phase of world commerce of which we are on the threshold, science, brain-power, energy, and faith must, and increasingly will, be harnessed to the work of perfecting air communication so that human mobility can be increased, knowledge interchanged, and the fruits of production distributed throughout the world.

As a soldier I have of course dwelt on the possibility of war in the future and of the part which aviation would play in it, but it would be a great mistake-- though I think that mistake is constantly made--to suppose that soldiers look forward with equanimity to the prospect of war. On the contrary, soldiers, more even than civilians, if this be possible, realize the horrors of war and recognize that the great task rests upon the statesmen of all nations, and upon humanity itself, of taking whatever steps can be taken to prevent its recurrence.

We may at least assume that another great war will not be allowed in our generation. But war, in spite of its horrors, in spite of its bereavements, is only too quickly forgotten. A comparatively few years, and those who have passed through its fire are no more. New wealth is created; new antagonisms arise; and a new generation remembers only the romantic stories and the martial deeds of its fathers, or, more fatally, organizes itself to avenge defeat. Then, once again, forgetful of the terrible lesson we have learned, the great nations of the world may unsheathe the sword as the only solution to their problems. Our only hope lies in using the ensuing years to educate mankind to the principle that war brings misery and impoverishment to all engaged in it, that in the final victory it is not a question of which is left the strongest, but which is the least exhausted, and that national are as susceptible as personal differences to discussion and arbitration. Above all, let us guard against the old mistake of competitive armaments. There is no reason, for instance, why, because France, our friend and ally, is adopting a policy of air armaments, we should blindly pile up aeroplane against aeroplane, pilot against pilot, and thus provoke mutual suspicion.

The possibility of war remains, however, and I wish in conclusion to emphasize the fact that in my belief the security of this country in the event of war will depend upon our strength in the air. The development of the offensive powers of aviation have already destroyed "the silver streak" on which we relied in the past. When we remember that it is less than twenty years since the first successful aeroplane was flown, when we recall the almost miraculous development of the fighting powers of aircraft during the four and a half years of war, and also the further developments which were on the point of being utilized when the war ended, it seems certain that from the point of view of war Britain has ceased to be an island. The "silver streak" would have been little protection but for our naval supremacy, and in the future our security will depend as much upon superiority in the air as it has depended in the past upon our superiority at sea. And this superiority in the air can only be attained in the same way in which we secured our supremacy at sea. That supremacy was not really gained by developing great navies. It was gained by our mercantile marine, which made the great navies possible. Our future security can only be gained by the development of commercial aviation.

Printed in Great Britain by Butler & Tanner, Frome and London.